The Implications of Colonizing Agriculture

Shashi Shekhar Prasad

RANDOM PUBLICATIONS
NEW DELHI (INDIA)

The Implications of Colonizing Agriculture

ISBN 978-93-5111-774-2

Published in 2016 in India by
RANDOM PUBLICATIONS
4376-A/4B, Gali Murari Lal, Ansari Road
New Delhi-110 002
Phone : +9111-43580356, 011-23289044, 011-43142548
e-mail: sales@randompublications.com,
info@randompublications.com, randomexports@gmail.com
Reprinted 2019

Type Setting by : Friends Media, Delhi-110089
Digitally Printed at : Replika Press Pvt. Ltd.

Preface

Indian agriculture began by 9000 BCE as a result of early cultivation of plants, and domestication of crops and animals. Settled life soon followed with implements and techniques being developed for agriculture. Double monsoons led to two harvests being reaped in one year. Indian products soon reached the world via existing trading networks and foreign crops were introduced to India. Plants and animals—considered essential to their survival by the Indians—came to be worshiped and venerated.

The middle ages saw irrigation channels reach a new level of sophistication in India and Indian crops affecting the economies of other regions of the world under Islamic patronage. Land and water management systems were developed with an aim of providing uniform growth. Despite some stagnation during the later modern era the independent Republic of India was able to develop a comprehensive agricultural programme.

The British agricultural revolution describes a period of agricultural development in Britain between the 16th century and the mid-19th century, which saw a massive increase in agricultural productivity and net output. This in turn supported unprecedented population growth, freeing up a significant percentage of the workforce, and thereby helped drive the Industrial Revolution. How this came about is not entirely clear. In recent decades, historians cited four key changes in agricultural practices, enclosure, mechanization, four-field crop rotation, and selective breeding, and gave credit to a relatively few individuals.

The challenges and issues of industrial agriculture for global and local society, for the industrial agriculture industry, for the individual industrial agriculture farm, and for animal rights include the costs and benefits of both current practices and proposed changes to those practices. Current industrial agriculture practices are temporarily increasing the carrying capacity of the Earth for humans while slowly destroying the long term carrying capacity of the earth for humans necessitating a shift to a sustainable agriculture form of industrial agriculture.

This book will be of immense value of students, teachers, teacher education, educational planner policy makers, curriculum formers agricultural administration and all for research scholars.

– ***Author***

Contents

1

Introduction

AGRICULTURE AND COLONIALISM

Over 2500 years ago, Indian farmers had discovered and begun farming many spices and sugar cane. It was in India, between the sixth and fourth centuries BC, that the Persians, followed by the Greeks, discovered the famous "reeds that produce honey without bees" being grown. These were locally called saccharum. On their return journey, the Macedonian soldiers carried the "honey bearing reeds," thus spreading sugar and sugarcane agriculture. People in India had also invented, by about 500 BC, the process to produce sugar crystals. In the local language, these crystals were called khanda, which is the source of the word candy.

Prior to 18th century, farming of sugar cane was largely confined to India. A few trader began to trade in sugar - a luxury and acostly spice in Europe until the 18th century. Sugar became widely popular in 18th-century Europe, then graduated to becoming a human necessity in the 19th century all over the world. This evolution of taste and demand for sugar as an essential food ingredient unleashed major changes. Sugarcane does not grow in cold, frost-prone climate; therefore, tropical and semitropical colonies were sought. Sugar cane plantations, just like cotton farms, became a major driver of large and forced human migrations in the 19th century and early 20th century - of people from Africa and from India, both in millions - influencing the ethnic mix, political conflicts and cultural evolution of various Caribbean, South American, Indian Ocean and Pacific island nations. The history and past accomplishments of Indian agriculture thus influenced, in part, colonialism, first slavery and then slavery-like indentured labour practices in the new world, Caribbean wars and the world history in 18th and 19th centuries.

IMPACT OF COLONIAL RULE OF BRITISH ON INDIAN AGRICULTURE

Farmingis the mainstay of Indian economy. Nearly eighty percent public adopted cultivation either as principal or as a secondary livelihood. About seventy percent of national earnings came from farm sector. Farm productions constituted mainly food-grains and such other harvests like oilseeds, fibre harvests, sugar cane required for domestic consumption. Moreover, agriculture had special importance in self-reliantrural economy. However, the British Rule altered the nature and structure of the Indian economy.

The land was heavily assessed for revenue; a new class of zamindars emerged; deindustrialisation led to overcrowding of land; increasing rural indebtedness put the farmers in poverty; a large number of intermediaries caused low productivity and finally the impoverishment of the peasantry were accelerated.

Under these circumstances, Indian agriculture could not sustain the pressure from the growing dependence on the land, the increasing Government dues and the injustice of unscrupulous zamindars. The consequence was unavoidable. Farming became stagnant and personality acre yields declined.

There were various factors contributing to stagnation of farming. It began with the land revenue policy of the Company. Ownership of land was vested with non-farmers where as the actual farmers had no claim over land. The Government became the rent receiver; the Zamindars were rent-collectors; and the farmers were mere rent payers.

The Government did nothing for farm development. The rent-collecting Zamindars had no interest in farming. Finally, the farmers had no resources for investment to improve farming. Moreover, the farmers lost interest to bring about progress in the land which they did not possess.

The land cultivated by him was not his property and the benefit coming out of farmingprogress would be reaped by the absentee zamindars and financiers. To them, farm progress meant the payment of more rent and no cultivator came forward to invest in fear of extra payment. Thus, farming declined steadily.

India handiwork industries were closed down and local markets were no more beneficial to the Indian traders. Within short-time, farming was left as the lone source of employment and thus got overcrowded due to migration of working persons from non -farm sectors. Further, uncontrolled population growth added extra pressure on land.

Thus, people competed among themselves for a plot of land and were exploited by rack-renting of the zamindars. The method of subletting the right to collect revenue created a chain of intermediaries and led to the subdivision and the fragmentation of land into smallholdings. As a result per capita land was very low and earnings from land could not meet the livelihood of the farmers. All apart, everyone wanted to be a rent collector instead of being a cultivator for which subletting and subleases increased. Thus, fragmentation of land into small holdings and excessive overcrowding reduced yields per acre. Indian farmers adopted primitive techniques in farm production. They hardly used better cattle and seeds, more manure and fertilizer and improved techniques of production. As discussed earlier, the farmers had no resource for progress of farming. The Government intentionally neglected farming.

Though the farmers shouldered the main burden of taxation, very small part of their taxes was paid for transformation of farming. The administration spent millions of rupees on the railways to promote the British trade interests. On the other hand, very little sum was spent on irrigation and that was the only field of Government investment.

The zamindars took no personal interest beyond the collection of tax. They exploited the farmers by rack-renting to enhance their earnings and were unwilling to make any investment to increase earnings by increasing output of land. Thus, farming continued to be neglected grossly and stagnation of farming was unavoidable.

No less were the effects of the natural calamities like floods, droughts and famines. The repeated occurrence of those calamities forced the farmers to give upon cultivation. There was no attempt to bring about any preventive measures against the natural calamities.

During the early years of the British Rule not anything was done to check or to regulate the flood water. No initiative was taken for providing irrigation that could have insured farm production against droughts or insufficient rainfall. Failure of harvests for two or more successive years took the dreadful shape of famine.

Neither the administration nor the zamindars paid any attention to prevent the devastation of the natural calamities. In India a good harvest depended on a good monsoon with adequate was uncertain, rainfall was irregular and natural calamities were inherent. The Government was apathetic, the zamindars were oppressive and the farmers were hopeless. Therefore, farming was left at the mercy of nature.

Similarly, no progress came in the farm technology. Farm implements were ordinary and old. Wooden ploughs were primarily used and cattle wastes constituted the manure. Use of iron ploughs was rare and an inorganic fertilizer was unknown. There was very little effort for creating educational awareness among scientific advancement would have been an effective measure to increase productivity. But the scientific stagnation fastened the decline in farming and ultimately poverty was perpetuated for rural masses more specifically for the farmers.

LAND REFORMS DURING 1947-70

A huge disparity in land holding pattern continues to exist in the country even after four decades of independence. A series of farm land reform laws have been passed by the State and Central Governments. Though the laws are very rosy and catchy but their implementation is hope- less. However, people working with rural poor masses of India must have an elementary knowledge about the existing farmland reform measures, operating in 'the country.

The stigma of Indian Farming is the highly defective structure of its farm land holdings. The measures of farm land reforms aim at correcting it. The term 'land reforms' involves procurement and redistribution of large holdings of farm land among the small farmers and landless farm labourers.

It is an instrument to bring about progress in the institutional framework of farm land. The responsibility of land reforms is owned by the government with a view of benefiting those who either have petty holdings or have no farmland at all. As big farm land owners are quite unlikely to share their holdings with their landless counterparts, intervention by the government using force of law/legislation is necessary to secure social justice for the masses.

LAND TENURE

Farmland tenure may be defined as the method in which an individual or the actual tiller of the land holds land; it determines his rights and responsibilities in connection with his holding. Obviously, the land tenure method refers to law/rules and regulations, which confer ownership rights upon an individual or actual tiller of the soil.

It determines the status of the actual tiller of the farm land and his relations with the state. If the actual tiller is not the owner of the farmland it determines the relation between the owner and the actual tiller of the farm land. It points out under what circumstances; the actual owner of the farmland may lose his own right. It specifies rent to be realised from the tiller, its time and methods. It specifies the conditions under which the actual tiller can sell or transfer his holding. It specifies the conditions, whether a cultivator cannot- gage his farmland or not.

There were a large number of land tenure methods prevalent in India in the pre - independence period. But the following three were more prevalent in different parts of the country.

Raiyatwari Method

Under this method, every registered holder is recognised as its owner. The owner cultivator or peasant proprietor is responsible directly to the government for the payment of farmland revenues and other dues. There is no intermediary between the government and the cultivator. This is perhaps the best method offarmland tenure. The peasant proprietor does not fear ejection by the government so long as he pays the farmland revenue. He can make permanent investments in his farmland as he is sure to reap its benefits. Thus, this method can ensure an increase in farm productivity.

Mahalwari Method

Under this method,farm land is held (owned) jointly by a collective body of the village. This body collects farmland revenues from the owners or cultivator farmers and is responsible to the government. This method is found in some parts of U.P., Punjab and Haryana. This method facilitates cooperative farming to get maximum yield from the land. The smallholdings of peasant farmers can be combined for this purpose. The main drawback with this method is that it encourages absentee landlordism.

ZAMINDARI METHOD

In Zamindari method, there is a separation of ownership of farm land from its farmers. Under this method, one person known as zamindar owns a village and is responsible for the payment of farmland revenues to the government. This method existed in West Bengal, some parts of U.P., 1 Maharashtra and Tamil Nadu. Now this method has been abolished.

THE NEED FOR LAND REFORMS

The defects existing with the Indian agrarian structure pointed out by the Planning Commission, highlighted the need for land reforms. The existing method during the beginning of Planned Growth, allowed the landlord and intermediaries to grow richer and they continued to flourish at the cost of the actual tillers. The cultivator tenants had to live a very tough life. Tenant got little incentive to increase his output for a large share went to the landowner.

Very small margin was left for the actual cultivator and this amount was quite insufficient to provide for a capital investment in the farmland. The zamindars grew richer, the intermediaries continued to flourish, the state was deprived of its share of legitimate increase in revenue and the cultivator tenants were in hand to mouth existence.

In order to remove the defects with existing agrarian structure, there was need of institutional changes in holdings. A high powered committee in 1948 with J. L. Nehru as its Chairman recommended that "all intermediaries between the tiller and the state should be eliminated and all middlemen should be replaced by non-profit making agencies like cooperatives. The maximum size of holdings should be fixed and the surplus farmland should be acquired and placed at the disposal of the village cooperatives. Small holdings should be consolidated and steps should be taken to prevent further fragmentation".

PURPOSE OF LAND REFORMS

The basic purpose of farmland reforms in India has been the creation of a method of peasant proprietorship. 'Land to the tiller' has been the motto. Through the redistribution of farmland by applying ceiling on farmland holdings, the idea has been to build up a vigorous independent peasantry consisting of small farmers and to help these farmers class with extension of credit and distribution facilities, largely through a network of cooperative service organisation.

The planners set out the objectives of the land reform policy as "the removal of such institutional and motivational impediments to the transformation of farming as were innate in the agrarian structure inherited from the past and the reduction of gross inequalities in the agrarian economy and rural society which stemmed from unequal rights in land".

The Planning Commission gave two basic objectives of land reforms, namely.

Economic Efficiency

The agrarian reforms should help in removing all obstacles to achieve higherfarm productivity. They should help in creating conditions for evolving as speedy as possible, afarm economy with a high level of efficiency.

Social Justice

The agrarian reforms should help to eliminate all elements of injustice and ensure social justice within the agrarian method to provide security for the tiller of the soil and assure equality of status and opportunity to all the sections of the rural population. *In order to achieve these objectives, the following policy measures were envisaged:*

- Abolition of the prevalent intermediary method between the state and the actual tillers;
- Tenancy reforms such as conferring of ownership rights on the cultivating tenants in the farmland held under their possession;
- Imposition of a ceiling on farmfarmland holdings as a measure contributing to the transformation of farming and to eliminate parasitic absentee farmlandlords;
- Rationalisation of the record of rights in farmland so as to make the rights of tenants, share harvesters and other categories of insecure zamindars;
- Consolidation of holdings with a view to making easier the application of modern techniques of farming; and Development of co-operative farming and co- operative village management.

PROGRESS OF LAND REFORM MEASURES

The farmland tenure method which the British imposed in India, regardless of the different juridical forms they assumed in different regions, where only variants of feudal and semi-feudal farmland ownership. The British administrators altered these methods in a manner as to facilitate the extraction of more rants from the farmers by making the landlord, who was earlier a rent collector, the absolute owner of farmland and by depriving the actual farmers of all their traditional rights. This was done in a forthright manner under the zamindarimethod and in yield and indirect manner in the Raiyatwari method. Although juridical no landlord or intermediaries were created and the settlements were made directly with the raiyat, yet the fact was that, due to prevailing inequalities in farmland holdings, the bigger raiyat zamindars came to dominate the agrarian set up in many respects and became the counter- parts of the zamindars in zamindari areas. Like the latter, they indulged in many semi-feudal forms of injustice such as share harvesting, rack renting, ejects, forced labour, usury, etc.

Under British rule, farmland reforms had a very limited scope and content. These farmland reforms were motivated not by consideration of improving production, nor did they have any sense of social justice. They were meant to safeguard British political influence in the rural areas and save the rural market from being completely pauperised.

Thus, the structure of agrarian society evolved under British rule, created a socioeconomic set up in which parasitism flourished, farmlandconcentrating in the hands of a few rural rich continued to grow, and landlessness and land hunger of the farmers mounted at an ever increasing pace. Evictions and insecurity of tenancy and rack- renting became a general phenomenon and the farmers were ground down by a colossal burden of indebtedness.

Farmland reforms have been on the agenda of rural reconstruction since independence. The state and central government have made the number of farmland reform laws after independence.

The reforms have been undertaken along the following lines:

- Abolition of zamindars and other intermediaries (jagirdars, inamdars, malgujars, etc.) Between the/state and the cultivator;
- Tenancy reforms and the reconstruction of the farmland ownership method;
- Fixation of ceiling on holdings and distribution of surplus farmland among the landless;
- Reorganisation of farming through consolidation of holding and prevention of further fragmentation; and
- Development of co-operative farming and co-operative village management methods.

Abolition of Zamindari Method

Intermediary tenures like zamindaris, joggers, imams, etc. Which prevailed over nearly 40 per cent of the cultivated farmland have been abolished. These intermediaries were responsible for the payment of farmland revenues to the government. The zamindars and other intermediaries were merely rent receivers and were not bothered about the progresss.

Reasons for Abolition

The general compulsion underlying the abolition of the intermediaries was the concentration of farmland ownership in the hands of a parasitic class who played no positive role in production while the vast mass of small farmers, who were the actual farmers, were divorced from the ownership of farmland. This discrepancy became the root cause of the state of chronic crises in which the Indian farm economy was enmeshed for several decades before the attainment of independence. It remained a completely stagnant economy. The rate of growth was less than half per cent. It was not expected that an utterly weak and unsuitablefarming of this nature would meet the growing demands of food and raw material of a new developing independent nation's economy. Thus the abolition of feudal and semi-feudal vested interests became an essential step for facilitating the growth of productive forces in the country. Besides, the intermediary method was subject to high rent and rack-renting, share-cropping, injustice and demoralisation of the actual tiller of the soil.

Thus, immediately after independence a strong voice was raised against the vested interests in farmland. As a result, top priority was given to the abolition of zamindari method or the intermediary method as a whole. Accordingly, every state enacted its own legislations for the abolition of intermediary interests. Legislative measures for the abolition of intermediaries were initiated soon after the independence, starting with Uttar Pradesh and being followed up in other states. The whole process of legal enactments on this issue was completed in the country within a decade, *i.e.* From 1950 to 1960. Since farmland reform was a state subject, actual enactments abolishing intermediaries were marked by certain variations from state to state though the salient features of most of those enactments were common.

Economic and Social Effect

Though the abolition of intermediaries was associated with many advantages, it had the following economic and social undesired consequences.

Heavy Burden of Compensation

The compensation to be paid to the intermediaries amounted to ₹ 670 crores as compared to additional farmland revenues of only ₹ 29.52 crores. This was a heavy burden on state governments. The compensation received by the zamindars was a waste of capital resources by the state governments. This amount was not invested in farming but in urban property or was spent in buying consumer durables.

Increase in the Number of Feudal Zamindars

The acts abolishing intermediaries did not divest the feudal zamindars of their holdings. This is because owners were allowed to retain as much farmland as they themselves can cultivate. Big farmland owners started cultivating farmlands by employing hired labour in any form. Thus, the "feudal landlord" was permitted or rather encouraged to shift! Their position from rent receivers ofself-farmers. It also encouraged "absentee landlordism" in which many intermediaries and non-farmers of soil became owners of farmland. They retained as many areas as they could show under their personal cultivation. Before the implementation of the act, the zamindars got the records manipulated into their favour in connivance with the local bureaucrats.

Eviction on a Large Scale

Since the intermediaries wanted to become owners of farmland under the garb of self-farmers, they evicted tenant farmers before the implementation of the legislation. The poor evicted tenants did not approach the court or the government because of poverty, fear and lack of organisation. Faulty records: The revenue records were faulty because of which many farmers could not get the benefit under the legislation of abolition of intermediaries.

Tenancy Reforms

The first phase of land reforms (1948-55) was mainly concerned with the abolition of intermediaries. The tenancy reform, which is the integral part of the land reform policy,favoured neither wholesale removal of landlordism nor the wholesale abolition of tenant farmers. Hence, the middle course was adopted. Thus, certain amendments to the existing tenancy laws were carried out along with the legislations for the abolition of intermediaries. This extended the scope of protection to the tenants of intermediaries particularly in areas of statutory landlord- deism.

But the owners were allowed to resume farmland for their personal cultivation. This led to the mass eviction of tenants, sub-tenants and share-croppers through various legal and extra-legal actions.In fact, a big drive to clear farmland of tenant occupants was started by zamindars in order to obtain maximum areas. Innumerable evictions were effected III the process of resumption of farmland by landowners. However, such evictions could not take place in the U.P. and Union Land of Delhi. In fact, U.P. has the credit of having the best land reforms in India.

To counteract this, the lawmakers in most of the states tried to enact or amend tenancy laws in the following decade (1955-65) and friends plug certain glaring loopholes in the existing laws. The major aspects incorporated in tenancy legislation in different states to protect the tenants can be identified as follows.

Fixation of Rents

Before the initiation of farmlandreform measures, the tenants were required to pay one half of their produce or more as rent to the zamindars.

During the first five-yearplan period, it was suggested that the rent should not exceed 1/4th or 1/5 of the produce in any case.

During the second and third plans also this suggestion was repeated and it was suggested that the rent should be made payable in cash. Legislation along these lines has been enacted in all the states. However; different states have prescribed different rates of rents. For instance in Gujarat and Maharashtra the maximum rent stands at one-sixth of the produce. In Assam, Manipur and Tripura maxi: mum rents vary between 1/4 to 1/5th of the gross produce. In Orissa and Bihar, 1/4th of the gross produce has been fixed as rent. In Rajasthan, fair rent is fixed at 1/6th of the gross produce but in case of cash rents, at twice the ' land revenue assessment.

Security of Tenants

It was emphasized in the first, second and third five year plans that the tenants. Should be accorded permanent rights In the farmlands leased In by them subject to a limited right of resumption to be granted to landowners. In accordance with this legislation, providing for security of tenure has been enacted in all states.

This legislation has three aims:

- The ejectment of the tenant should not take place except in accordance with the provision of law;
- The land may be resumed by the owner only for personal cultivation; and
- In the event of resumption, the tenant is to be assured of a minimum tenanted area in his possession.

The legislation provided that the ejectment of tenant can take place only through order of a revenue court.

The grounds on which ejectment may be allowed include:

- Non-payment of rent,
- Performance of an act which is destructive or permanently injurious to the land,
- Sub- letting the land, and
- Resumption of land for personal cultivation by the landlord.

Legislations have been passed for granting security of tenure in different states on the following patterns.

- All tenants in possession of cultivated farmland have been given full security of tenure. The landowners have no right to resume farmland for personal cultivation as in V.P. and Delhi.
- In Assam, Maharashtra, Gujarat, Punjab, Rajasthan and Himachal Pradesh landowners are permitted to resume a limited area for personal cultivation subject to the condition that a minimum area or portion of the holding is left with the tenants.

- In West Bengal and Jammu and Kashmir, a limit has been placed on the extent of farmland, which a landowner may resume. But the tenant is not entitled to retain a minimum area or portion of this holding in all cases.

Right of ownership

Regulation of rents and security of tenure is treated as the first stage in the tenancy reforms. The ultimate goal is to confer rights of ownership on as many tenants as possible and bring them in direct contact with the state. Legislations passed along these lines provide for bringing tenants of non-resumable farmlands into direct relationship with the state in the following three ways.

- By declaring tenants as owners; the tenants were required to pay compensation to owners in suitable installments;
- Through the acquisition of right of ownership by the state on payment of compensation and transfer of ownership to tenants; and
- By protecting the interests of sub-tenants under the tenancy laws and bringing them into a direct relation - ship with the state.

The impact of these measures can be seen from the pattern of holding that has now emerged in the country. The farm census report pointed out that out of the 70 million holdings in the country 64 million or 92 per cent holdings are wholly owned or self-operated, 3 million holdings are partly owned and partly rented and another 3 million holdings are wholly leased accounting for 4 per cent each. Out of 162 million hectares under the holdings, 148 million hectares (91 per cent) are wholly owned and self operated; 10 million hectares (6 per cent) are partly owned and partly rented and the balance 4 million (3 per cent) hectares are wholly leased. It is now obvious that the most of the holdings are now under self-cultivation and the evil of share-croppers has been reduced to a great extent.

It is because of high re- turns into self-cultivation and the owner cultivator do not find it beneficial to lease out farmland on share harvesting. In view of the increasing pressure of population on farmland and unemployment the leasing out of farmland is expected to be a rare phenomenon in future. Those farmers' who do not possess the required amount of labour and capitals adopt the practice of leasing out of farmland. Otherwise, in view of high returns from farmland, leasing out and share-cropping are considered unbeneficial by owner farmers.

EVALUATION OF TENANCY REFORMS

Because of tenancy legislations in many states the tenants and sub-tenants have been brought into direct relationship with the state. But the progress was very slow in some states due to the following reasons. -The legislation has not been able to meet the objects laid down by the Planning Commission.

The fixation of statutory rent was very high in some states:

- The term personal cultivation was defined in a loose manner. Because of this, the farmlands ostensibly resumed by the landlords on the pretext of personal cultivation are cultivated through harvest -sharing arrangements where the sharers are treated as labourers.
- The definition of the term 'tenant' excluded the share-croppers who form a vast majority of the tenant farmers. Thus share-croppers did not get any benefit.
- The non-availability of correct and up-to-date farmland'. Records have not allowed to carry out the tenancy reforms properly.
- The tenants can be evicted from their holdings on many grounds. This has been termed as a continuing hang over of the feudal method.

Suggestions for Progress

The following suggestions can be considered useful in achieving the aims of tenancy reforms:

1. The resumption of farmland by the owner should be legal in cases where the owner cultivates the farmland himself.
2. The ex-zamindars who have retained excess holdings under the pretext of personal cultivation should be brought under a ceiling limit.
3. Correct farmland records should be compiled and maintained so as to facilitate effective application of tenancy farmland reforms.
4. There should be a complete ban on letting and subletting of farm lands. An Exception should be allowed in cases where the owner is a widow, minor or handicapped. The real purpose of land reforms can be served only if the farmers get financial support from the financial institutions for the permanent progress of their holdings.

Ceiling on Land Holdings

Ceiling on farmland holdings refers to the fixation of maximum size of a holding that an individual cultivator or a household may possess. Beyond this maximum limit, all farmland belonging to the zamindars is taken away by the government to be redistributed among the landless labourers. Thus the imposition of a ceiling on farm holdings is mainly a redistribution measure.

The idea basically is to ration the farmland in such a way that, above a certain level, the surplus farmland is taken away from the pre- sent holders and is distributed to the landless or to the small farmers. This will reduce the wide disparities of earnings and wealth found in the agrarian structure.

The ceiling question gave rise to more debate and arguments than any other reform issue. Legislations for ceiling on existing holding and future acquisition were enacted in most of the states during the second plan period.

The ceiling on farm holdings was intended to:

- Meet the hunger of the landless;
- Reduce the glaring inequalities in farmland ownership so as to pave the way for development of co-operative rural economy; and
- Increase self-employment in owned farmland.

The ceiling laws were enacted and enforced actually in two phases: the earlier phase covering the period from 1960-1972, before the national guidelines were laid down, and the latter since 1972 after the adoption of the guidelines. However, provisions related to ceiling laws can be analysed under the following heads.

Unit of Application

In the beginning some states took 'individual' as the unit of the ceiling, while some others regarded 'family' as the unit. This led to widespread irregularities and big farmland owners started transferring their farmland into pieces to their fake kiths and kins and managed to keep unduly large holdings. However, since 1972, after suitable corrections, the unit of farmland ceiling universally adopted by all the states is a family having a father, a mother and children. Parents having more than 5 children can be given a little exemption but in no way the amount of exemption will exceed twice the prescribed limits.

Level of Ceiling or Maximum Limit

Under the old acts (prior to 1972) there were wide variations in the fixation of ceiling on farmland holdings since the ceiling was a state subject and each state enacted its own ceiling laws. For instance the limit of ceiling prior to 1972 in Andhra Pradesh varied from 27 to 234 acres, in Assam it was fixed at 50 acres, in Gujarat it varied between 10 to 132 acres, Haryana 30-60 standard acres, Kerala 15 acres, M.P. 25 acres, Orissa 20-80 acres, Punjab 30-40, acres, U.P. 40-80 acres, Tamil Nadu 30 acres to 120 ordinary acres, and West Bengal 25 acres.

Thus, there was a large gap between minimum and maximum and from state to state permissible limits of farmland holdings. As such, much farmland could not be achieved for redistribution. Since 1972 these limits have been rationalised. Thus under the new policy, the upper limit of the ceiling has been lowered and the range between the lowest and the highest has been considerably narrowed. For irrigated farmland where at least two harvests are raised, the ceiling, depending on the productivity of farmland and other factors, has been fixed at 10-18 acres. In places where irrigation is done by private sources, for the purpose of fixation of ceiling, 1.25 acres is deemed to be equivalent to one acre of farmland irrigated by public sources. However, in both cases, the upper, limit does not exceed 18 acres. In areas where there is a provision for irrigation for the raising of only one harvest, the upper limit of the ceiling has been fixed at 27 acres. For, remaining types of farmland which are not so productive, the I~ upper limit. Is 54 acres.

Exemptions

There was a long list of other kinds of farmlands whichwas exempted from the purview of ceiling legislation before 1972. For instance, tea, coffee, and rubber plantations, orchards, specialised farms engaged in cattle breeding, dairying, wood raising etc., Sugar cane farms, factories and efficiently managed farms which consist of compact blocks on which heavy investment or permanent structural progress have been made were exempted. In those parts where cultivable wastes are available and sufficientnumber of farmers is not forthcoming, the ceiling may not be imposed or may be placed at a higher level.

The farmland ceiling measures have been disappointing in the country except in the states of J and K and West Bengal. By the end of 1970 about 11 lakh hectares of farmland could really be taken over and again only a half of the taken- over farmland could actually be distributed among the landless labourers. In some states such as Bihar, Karnataka, Orissa and Rajasthan, no farmlandwere declared surplus. The situation did not improve even after 1970.

Recommendations of the Chief Minister's Conference

In 1972, a new national policy on the land ceiling was evolved on the recommendations of the chief minister's conference. The following guidelines were laid down.

- Taking into account the fertility of soil and other conditions, the best category of land in the state with assured irrigation and capable of yielding at least two harvests a year the ceiling should be fixed 10-18 acres.
- In the case of land with assured irrigation for only one harvest in a year, the ceiling shall not exceed 27 acres.
- In case of owners with different types of holdings, after converting the better categories of land into lowest categories, the ceiling should not exceed 54 acres.
- The unit of application should be a family of 5 members. Where the members of the family exceed additional land may be allowed for each member in excess of 5 in such a manner that the total area admissible to the family does not exceed twice the ceiling limit for a family of 5 members. Where both husband and wife hold land in their own names the two will have rights in the properties within the ceiling in proportion to the value of land held by each before the application of the ceiling. Each major child is to be f treated as a separate unit for the purpose of application of the ceiling.

Consolidation of Holding

India is a land of very small, fragmented and uneconomical holdings. That is why the need of consolidating these fragmented holdings was felt in order to improve their productivity and viability of investments. Various state governments, on these lines, have introduced legislations. Consolidation of farmland is a process of rearrangement of farmland on the basis of existing rights. Most states have not shown any enthusiasm for implementing such legislations. Only in Punjab, Haryana and parts of UP. Thisprogramme has made desired progress. Orissa, Bihar, H.P. etc. Have also taken up consolidation in a big way. An area of 584.72 lath hectares has so far been consolidated all over India.

IMPACT OF LAND REFORMS

The impact of farmland reform measures on the agrarian structure of the country can be discussed under the following heads.

Changing Over to Market Oriented Farming

The analysis of pre-independence patterns of farmlandmethod reveal that the agrarian and social structure which developed under the British tended to perpetuate a back- ward and medieval economy in a state of stagnation for decades. The forces to impede the production and development were very active. In contrast to this, National Commission on Agriculture (NCA) pointed out that the essence of the present situation is that Indian farming is in a stage of transition from predominantly semi-feudal type of farming characterised by large scale leasing of farmland and subsistence farming to a commercialised farming and, thus, increasingly assuming the character of market oriented farming.

End of Feudalism

The National Commission on Agriculture also pointed out that as a result of farmland reforms that have taken place since the independence, the feudal and semi-feudal farmland owning classes have lost their erstwhile domination over the Indian agrarian economy as a whole. Moreover, the decline in the semi-feudal relations have followedthe growth of farming on commercial lines. However, some of the evils such as share-cropping, extraction of high rates of rents, usury, eviction of tenants, social and caste oppression, etc. Still prevail in some parts of the country but the degree of their intensity is negligible.

Leasing in of Land by Big Owners

One of the important effects of farmland reforms is that, the subsistence farming is changing over to commercial farming. In this context NCA has pointed out that under commercial tenancy, leasing in of farmland by the big farmland owners from the small farmers takes place and such tenancy prevails more in areas where farming is modernised. It is common in Punjab and other areas where the impact of the green revolution has been appreciable. It is almost absent in the eastern regions of the country where farming is far less developed and where the old type of tenancy still persists.

Emergence of Modern Entrepreneurs

As a result of land reforms a class of modem entrepreneur farmers has emerged. These farmers have substantial farmland holdings and cultivate their farmland through hired labour and new technology. They are drawn largely from the ranks of exfeudal zamindars, upper strata of privileged tenants and the bigger raiyots, financiers, f merchants and various other categories of substantial zamindars.

Besides the growth of commercial farming and the rise in the prices of farm commodities and also I progress in techniques, have strengthened the economic position of this class of big farmers. They are also the main beneficiaries of governmental expenditure on farm development. It is this class which has been treated as the main custodian of the 'green-revolution'.

Reduction of Poverty

Besides several negative impacts, land reform measures have certainly reduced the disparities in farm holdings. The surplus farmlands of big zamindars have been distributed among the tenants and small farmers. The injusticeof tenants by the farmland owners has been reduced considerably. The cultivator-owner has been given assistance by the credit institutions to increase the productivity of their farmlands. The cultivator-owner has been brought in direct contact with the state. They are no longer required to share their produce with their zamindars. All these steps have led to an increase in the earnings of the small farmers and thus reduced poverty in the rural areas.

Use of Institutional Credit

Agrarian reforms have significant implications in facilitating the use of institutional credit. The farmland reform measures have influenced the working of financial institutions *viz.* Co-operative banks, regional rural banks and commercial banks, etc.

Review of Land Reform Measures

Though the farmland reform measures have been instrumental in bringing about some desirable changes in Indian agrarian structure, yet, they have failed to secure a justice to a large section of the rural population. The results of land reforms implemented so far have been far from satisfactory. Because of certain loopholes in the reform policies and implementation methodology, the progress has been very slow. In 1980.81 more than 54 per cent holdings were less than 1 hectare each.

Only 32 per cent holdings were in the range of 1 to 4 hectares. More than 70 per cent of the farmers still have to survive on less than 30 per cent of the total arable farmland of the country. These figures reveal the existing wide disparities in farm holdings despite 40 years of planned economic development.

Some of the glaring examples of weakness in farmland reform measures are listed as below:

- There has been no uniformity in execution of farmland reform policies and legislations in different states. For example, the rent to be charged from the farmers shows a wide variation from 20 per cent to 50 per cent of the gross product.
- Land reform measures have failed to prevent subletting and rack-renting.
- The identification of self-cultivation was wrong which allowed the big zamindars to keep large holdings with them.
- Ex-zamindars and zamindars have showed on papers that they have been cultivating the land. However, they get them tilled by hired labourers.
- Eviction of tenants has occurred on a large scale out of the suspicion of land being lost and under the guise of "voluntary surrenders".
- Administrative arrangements for enforcement and: supervision of land reforms have been fully inadequate.
- Records of tenants did not exist in several states and often incomplete and out of date records were used for the implementation purposes.
- The several states the existing provisions of security of tenure were of an interim nature and comprehensive measures to bring tenants into direct relation with the state are yet to be adopted.
- The rights to resumption widened the scope of ejectment.
- To provide security of tenure to the tiller, the landlord tenant bond was to be broken by the state interposing between them to collect fair rents from the tenant and pay it to the landlord after deducting the land revenue and collection charges.

REASONS FOR LOW PROGRESS OF LAND REFORMS

The task force on agrarian relations set up by the Planning Commission to appraise the progress and problems of farmland reforms, identified the following reasons for the poor performance of land reform measures.

Lack of Political Will

In the context of the socioeconomic conditions prevailing in the country, no tangible progress can be expected in the field of land reforms in the absence of requisite political will. The sad truth is that this crucial factor has been wanting. In no sphere of public activity in our country since independence has had such a big gap between precept and practice *i.e.* Between policy pronouncement and actual execution.

Absence of Pressure from Below

Except in a few scattered and localised pockets, practically all over the country, the poor farmers and farm workers are passive, unorganised and inarticulate. The basic difficulty in our situation arises from the fact that the beneficiaries of farmland reforms do not constitute a homogeneous social or economic group.

Negative Attitude of the Bureaucracy

Towards the implementation of farmland reforms, attitude of the bureaucracy has been generally lukewarm and indifferent. This is, of course, unavoidable because, as in the case of men who wield political power, those in the high echelons of the administration also are either big farmland- owners themselves or have a close nexus with big farmland- owners.

Legal Hurdles

Legal hurdles also stand in the way of farmland reforms. The task force categorically states: "in a society in which the entire weight of civil and criminal laws, judicial pronouncements and precedents. Administrative procedure and practice is thrown on the side of the existing social order based on the inviolability of the private property, an isolated law aiming at the restructuring of the property relation in the rural area has little chance of success. And whatever little chance of success was there, completely evaporated because of the loopholes in the laws and protracted legislations".

Absence of Correct and Up-to-Date Land Records

The absence of correct and complete farmland records further added a good deal of confusion. It is because of this that no amount of legislative measures could help the tenant in the court unless he could prove that he is the e actual tenant. This he could only do if there were reliable, and up-to-date records of tenants.

The main reason for the unsatisfactory state of affairs are (a) many of the areas in the country have never been cadastrally surveyed, (b) in some areas where cadastral surveys were done for a r long time, no resurveys have been taken, (c) no machinery, of any kind existed for maintaining village records, (d) even where records were kept by government officials, there is no uniform method, and (e) it has been found that even official records in many cases have not been correct.

Lack of Financial Support

Lack of financial support plagued the Farmland Reform Act from the beginning. No separate allocation of funds was made in the fifth plan for financing farm land reforms. Many states declined to include even expenditure of such essential items like preparation of records of rights in their plan budget. The state plans which are nothing but the aggregate of expenditure programmes hardly made any reference to farmland reforms. Whatever funds were needed for finalising of this programme had to be provided in non-plan budgets. It is because of this that the expenditure for farmland reforms was always postponed. Or kept to ' the minimum.

Land Reforms have Been Treated as an Administrative Issue

The implementation of farmland reforms is not an administrative issue, it is more of a political issue. Therefore, it is necessary to strengthen the political will for implementing farmland reforms. The task force of the Planning Commission in a very forthright manner states: "it should, however, be clearly understood that the mere setting up of an efficient administrative machinery will not by itself lead to any substantial progress unless the political and economic hurdles operating against the programme are removed."

PROGRESS

The farmland reform measures were thoroughly reviewed by the Planning Commission and other committees. Based on their findings, National Commission on Agriculture has suggested following measures to reinforce the implementation of farmland reforms.

Breaking of Landlord-tenant Nexus

The private owner who rents out farmland today is undoubtedly intermediary between the tiller and the state and, as such, does not fit into the fundamental pattern of farmland reforms. It is, therefore, high time that tenancy I reforms were directed towards the state of finally breaking up the farmlandlord-tenant nexus.

A potent practice of farming *i.e.* landlordism should be discouraged and ultimately curbed. Farming should be treated as a family occupation of the peasant-cultivator and not a source of subsidiary unearned earnings. In a normal peasant-proprietor economy absentee landlordism should find no place.

Restricted Tenancy Should be Allowed

Under the present farmland-man ratio existing in India, tenancy as such cannot be banned though undesirable. Experience has shown that wherever such an overall ban has been imposed, it has led to the emergence of concealed tenancy with all its evils. Hence, until such time as socioeconomic development in the country bring about a radical change in man-farmland ratio and create possibilities of transfer of population from the farming to non-farm sectors on a big scale, the tenancy shall have to be permitted in a restricted form. It will have to be strictly controlled and regulated.

All tenants of landowners excepting such landowners who possess land up to a marginal holding should be vested with proprietary rights and simultaneously declared virtual owners on a date to be specified by the state government. However, disabled persons, minors, widows and army personnel's are given some concession, the nature of which may be decided by the state governments.

This provision shall not apply to those cases where a bigger landowner has leased in land from a small landowner. Share-harvesters who have so far been not treated as tenants should be recognised and regarded as tenants and given all due protection as such.

Effective Implementation of Ceiling Laws

The studies have revealed that ceiling laws have not been able to make any appreciable breakthrough in reducing concentration of land in the hands of a few big farmers. These laws are devised to achieve the purpose of substantially reducing the present inequalities in farmland holdings.

It is, therefore, suggested that present ceiling legislations should be enforced vigorously. For instance, it is necessary to take firm measures against fiction and benami transfers which have been intentionally manipulated by big farmland owners in order to by-pass ceiling legislations.

The state government should conduct a proper enquiry into such transfers. If it is found that the transfers were made purposely to evade the provisions of ceiling laws, the farmland so transferred should be vested in the state after imposition of any penalty on the transfer.

In cases where fictions co-operative societies have been organised with a view to concealing the surplus farmland, such co-operatives should be subjected to proper investi- gation. And where many partners have been shown in a holding, but the holding as a whole is under a single management, such cases should also be brought to the lime- light and adequate actions should be taken to undo it.

Control on Land Held by Trusts and Institutions

The Planning Commission holds the view that the land possessed by trusts or institutions used for religions, charitable or educational purposes should be brought under ceiling laws. It would not be proper to give them a blanket exemption because a large number of such institutions and trusts are fictions or have been intentionally created to evade ceiling laws. Some of them are not being used for the purpose which on paper they profess to serve. Arable lands held by such trusts and endowments should be brought under the ceiling laws and more ceiling should also be made applicable to forests and water areas held by such institutions.

Distribution of Surplus Land

For distributing the surplus land accumulated out of enforcing ceiling legislations, opinions are different as to who should be preferred for allotment, landless labourers or small farmers owing less land than the minimum limit in order to make such holding viable. There is enough force in the argument that land should be distributed. To a small farmers owning less land than the economic holding or the minimum operational holding. Some have also argued that it is not only important to fix a ceiling on land holding, but it is also important to fix floor so that a large number of farmers may have at least a small operational holdings.

But in the view of massive landlessness, a serious lack of employment opportunities and a subsistence level of, almost half of the rural population below the poverty line, the land should be distributed to the landless labourer to whom land, however small, is the source of employment and relief from destitution. Hence for a long time to come a floor or ownership cannot be applied.

Simplification of Legal Procedure and Administrative Machinery

The Planning Commission has observed that ceiling legislations have suffered not only because of certain political and economic constraints but also because of a very inadequate and inefficient administrative machinery for enforcing it. The same is true for other kinds of land reform measures. It means that existing administrative machinery has generally failed to prevent the evasion of effected laws and has been functioning largely in collusion with the vested interests, especially in the village and tehsil levels.

The existing districts civil and revenue courts cannot properly discharge those functions being already over- burdened with other kinds of litigations. Besides, the existing method causes unnecessary delays and makes justice very costly. It often results in the dispensation of a doubtful nature. These courts are far away from villages and the poor man is generally at a disadvantage. Hence, the restructuring of administrative machinery is required.

Voluntary Surrender Should not be Accepted

Voluntary surrender have generally been used to cover up forcible and illegal eviction of tenants. Such surrenders should not be accepted as valid unless they are certified as genuine by an appropriate authority. In this context, the commission suggested that land surrendered should not revert to the landowner but should vest in the state to be allotted to any other eligible person. It can also be argued that it should revert to the owner if he possesses land less than the ceiling limit.

Higher Rent Should be Curbed

Despite the fixation of rent in most states on the lines recommended under various plans, higher rent still prevails in many parts of the country. It is, therefore, suggested that such rent receipts should be strictly enforced. The tenants should be entitled to remit their rents through financiers or deposit them in the tehsil.

Preparation of Up-to-Date Land Records

Tenancy legislation cannot be properly implemented without adequate and proper land records. Therefore, it is imperative that the preparation of land records should be given top priority in the whole scheme of enforcement of land reforms. Tenants, tenants-at will, and share- harvesters should be promptly and properly identified and their names should be recorded forthwith. It can, how- ever, be said that the interest of the owner should not be ignored.

Ensure Security of Tenure

So far as the tenure is concerned, the most important requirements would be:

- Recording the names of all the persons who hold land including share-harvesters in the record of right;
- Ensuring that not more than the legally stipulated share of the harvest is taken from the shore-harvesters by land owners;
- Ensuring that no ejectments takes place either on the basis of voluntary or through other extra-legal or illegal methods;
- A ensuring to the heirs of the share-harvesters on their death where the law provides it; and providing supportive services including credit to share-harvesters to free them from the clutches of landowners and financiers.

LAND USE AND OWNERSHIP

Out of 304 million hectares of land in India for which records are available, roughly 40 million hectares are considered unfit for vegetation as they are either in urban areas, occupied by roads and rivers, or under permanent snow, rock or desert.

Of the remaining 264 million hectares of land that have some potential for vegetation, 142 million hectares are cultivated, 67 million hectares are classified as forestland, and 55 million hectares as fallow or wasteland, or land with pastures or groves.

In percentage terms, according to World Bank estimates:

- Cultivable land amounts to around 58 per cent of land that has potential for vegetation.
- 22 per cent are forestland.
- 7 per cent are uncultivated (revenue) 'wasteland'.
- 7 per cent are rocky, barren land.
- 7 per cent is urban/non-farm land.

Roughly 20 per cent of the total land area are 'commons', which includes both cultivable and uncultivable wasteland and some forest land.

The Central Statistical Organisation puts the percentage distribution of the country's total land area of land use (1992-93 figures) as follows:

- Forests: 22.2 per cent
- Land not available for cultivation: 13.3 per cent
- Permanent pastures and other grazing land: 3.3 per cent
- Land under tree harvests included in net area sown: 1.2 per cent
- Cultivable waste land: 5.1 per cent
- Fallow land: 8.2 per cent
- Net area sown: 46.3 per cent
- Area sown more than once as percentage of net sown: 25.7 per cent

LAND OWNERSHIP IN INDIA

During the two centuries of British rule, India's traditional land ownership and land use patterns were changed. The concept of private property was introduced, de-legitimising community ownership methods in tribal societies.

The British introduced the 'zamindari' or 'permanent settlement method ' in 1793, whereby feudal lords became owners of large tracts of land against fixed revenue payments to the government. Farmers became tenant farmers and had to pay rent. This method prevailed in most of northern India.

In the south and west, the 'Riyatwari method ' was followed. Individual farmers (ryots or raiyats) were proprietors of land against revenue payments, with rights to sub-let, mortgage and transfer land. A third method under British rule was the 'mahalwarimethod ' whereby entire villages had to pay revenue, with farmers contributing their share in proportion to their holdings.

Land distribution under these methods became extremely unequal, and rural society got polarised into zamindars and rich farmers versus tenants and farmlaborers.

The land transfer was institutionalised under British rule and financiers secured land against loans. Combined with high revenue rates, this led to growing indebtedness, dispossession of land, rising tenancy, and a widening of the earnings gap between rich zamindars and poor tenants and farmlaborers. By Independence, about 40 per cent of India's rural population was working as landless farm labour.

Thus India has inherited a semi-feudal method of land distribution that followed the social hierarchy. Most landowners belong to the upper castes and farmers to the middle castes; farm labourers are largely Dalits and adivasis.

After Independence, India brought in legislation for land reform that included:

- Abolition of the zamindarimethod.
- Abolition of intermediaries.
- Protection to tenants.
- Rationalisation of different tenure methods.
- Imposition of ceilings on land holdings.

However, the legislation did not lead to substantial progress towards equitable land distribution. Most studies in fact show that inequalities have increased rather than decreased. The number of landless in India has progressively increased.

Landholding distribution too has become skewed.

According to government data compiled from sources such as the all India Report on Agriculture Census 1991-2000, in 1995-96:

- 1.2 per cent of landholdings in the country accounted for 14.8 per cent of the total operational holdings with large holdings of 10 hectares and above (average holding: 17.21 hectares).
- 6.1 per cent of holdings accounted for 25.3 per cent of the total operational holdings with medium holdings of 4 to 10 hectares (average holding: 5.8 hectares).
- 12.3 per cent of holdings accounted for 23.8 per cent of the total operational holdings with semi-medium holdings of 2 to 4 hectares (average holding: 2.73 hectares).
- 18.7 per cent of holdings accounted for 19.8 per cent of the total operational holdings with holdings of 1 to 2 hectares (average holding: 1.42 hectares).

As many as 61.2 per cent of holdings accounted for only 17.2 per cent of the total operational holdings. On average, the size of these marginal holdings was 0.4 hectares.

LANDLESS LABOUR

According to the India Rural Development Report of 1992, 43 per cent of the country's rural population were absolutely or near landless. Landless farm labour makes up almost half of those living below the poverty line in rural India. A majority of the economically and socially weaker sections of society, such as scheduled castes and tribes, Dalits, adivasis and women, make up the majority of landless population working as labour.

Landlessness has been steadily rising among the scheduled castes and scheduled tribes. According to a government Rural Labour Enquiry report, the percentage of landless households among scheduled castes increased from 56.8

per cent in 1977-98 to 61.5 per cent in 1983, while among adivasis it increased from 48.5 per cent in 1977-78 to 49.4 per cent in 1983.

Even among those who own land, a majority own marginal plot that provides them little or no food security. The government describes such marginal landowners as 'mere landless' (those who own less than 0.002 hectares) and 'near landless' (those who own between 0.002 and 0.2 hectares).

THE PRESENT HARVESTPING PATTERNS

As indicated earlier in this chapter, we can hardly describe all the harvesting patterns within the framework of this chapter. Therefore only important ones are highlighted. There are many ways in which a harvesting pattern can be discussed.

A broad picture of the major harvesting patterns in India can be presented by taking the major harvests into consideration. To begin with, the southwesterly monsoon harvests (kharif), Bajra, maize, Ragi, groundnut and cotton. Among the post-monsoon harvests (Rabi), wheat, sorghum (Rabi) and gram can also be considered to be the base harvests for describing the harvesting patterns. With such an approach, the harvest occupying the highest percentage of the sown area of the region is taken as the base harvest and all other possible alternative harvests which are sown in the region either as substitutes of the base harvest in the same season or as the harvests which fit in the rotation in the subsequent season, are considered in the pattern. Also these harvests have been identified as associating themselves with a particular type of agroclimate, and certain other minor harvests with similar requirements are grouped in one category. For example, wheat, barley and oats, are taken as one category. Similarly the minor millets (Paspalum, Setaria and Panicum spp.) Are grouped with sorghum or Bajra. Certain other harvests, such as the plantation harvests and other industrial harvests are discussed separately.

THE KHARIF-SEASON HARVESTPING PATTERNS

Among the kharifharvests, rice, jowar, Bajra, maize, groundnut and cotton are the prominent harvests to be considered the base harvests for describing the kharifharvesting patterns.

The rice-based harvesting patterns

Rice is grown in the high-rainfall area or in areas where supplemental irrigation is available to ensure good yields. If the harvest has to depend solely on rainfall, it requires not less than 30 cm per month of rainfall over the entire growing period. However, only 9 per cent of the area in the country comes under this category, and it lies in the eastern parts. Nearly 45 percent of the total rice area in India receives 30 cm per month of rainfall during at least two months (July and August) of the southwesterly monsoon and much less during other months. In contrast to these parts, the eastern and southern regions comprising Assam, West Bengal, coastal Orissa, coastal Andhra Pradesh, Karnataka (most part), Tamil Nadu and Kerala receive rainfall of 10 to 20 cm per month in four to eight consecutive months, starting earlier or going over later than the southwesterly monsoon months. With supplemental irrigation, 2 or 3 harvests are taken in these areas. However, it has been observed that on an all-India basis, nearly 80 per cent of rice is sown during June-September and the rest during the rest of the season. Areawise the mono-season belt occupies 53.6 per cent of the area (comprising Assam, West Bengal, coastal Orissa, coastal Andhra Pradesh, parts of Tamil Nadu, Karnataka and Kerala).

On an all-India basis, about 30 rice-based harvesting patterns have been identified in different states. In the most humid areas of eastern India comprising Tripura, Manipur and Mizoram, rice are the exclusive harvest. In Meghalaya, rice is alternated with cotton, vegetable and food-harvests, whereas in Arunachal Pradesh, where rice is not grown exclusively, the alternative harvests being maize, small millets and oilseeds. In parts of Assam, West Bengal, Bihar, Orissa and northern coastal districts of Andhra Pradesh, jute forms an important commercial harvest alternative to rice. In West Bengal, besides rice and jute, pulses and maize are grown on a limited scale. In Bihar, rice is grown over 49 per cent (5.3 m ha) of its harvested area (14.2 per cent of all-India area), whereas pulses, wheat, jute, maize, sugarcane and oilseeds are the alternative harvests.

In Uttar Pradesh rice is grown on 19 per cent (4.6 m ha) of its harvested area and represents about 12.4 per cent of the all-India are under this harvest. Rice is concentrated in the eastern districts of Uttar Pradesh where the alternative harvests are pulses, groundnut, sugarcane, Bajra and jowar in the decreasing order of their importance. Tobacco is grown in some districts.

In Orissa, rice is grown on more than 50 per cent of the area, whereas the alternative harvests are: pulses, Ragi, oilseeds, maize and small millets. In Madhya Pradesh rice is grown in the Chattisgarh area on 4.3 m ha (11.7 per cent of the all-India rice area), but the harvest suffers because of inadequate rainfall and irrigation. The important alternative harvests of this area are: small millets, pulses and groundnut. Wheat is also grown on a limited scale.

In the southern states, namely Andhra Pradesh, Tamil Nadu and Kerala rice are grown in more than one season and mostly under irrigation or under sufficient rainfall. Together, these three states have over 6.0 m ha, representing over 17 per cent of the all-India area under rice. Important alternative plantation harvests in Andhra Pradesh are: pulses, groundnut, jowar, maize, sugarcane and tobacco.

In Karnataka the harvests alternative to rice are: Ragi, plantation harvests, Bajra, cotton, groundnut, jowar and maize. In Kerala plantation harvests and tapioca form the main plantation harvests alternative to rice. In Maharashtra rice is grown mostly in the Konkan area over 1.3 m ha, along with Ragi, pulses, Rabijowar, sugarcane, groundnuts and oilseeds. In other states, namely Gujarat, Jammu and Kashmir, Rajasthan and Himachal Pradesh, rice form a minor plantation harvest and is mostly grown with irrigation. However, in Punjab and Haryana and to some extent in western Uttar Pradesh owing to the high water - table during this monsoon season, rice has become a major harvest in such areas.

The Kharif Cereals Other than Rice

Maize, jowar and bajra form the main kharif cereals, whereas ragi and small millets come next and are grown on a limited area. By and large, maize is a harvest grown commonly in high-rainfall areas, or on soils with a better capacity for retaining moisture, but with good drainage. Next comes jowar in the medium rainfall regions whereas Bajra has been the main harvest in areas with low or less dependable rainfall and on light textured soils. The extent of the area under these harvests during the south-westerly monsoon season is maize, 5.6 m ha; jowar (kharif), 11 mha, and bajra,12.4 m ha. Even though these harvests are spread all over the western, northern and southern India, the regions of these harvest patterns are demarcated well to the west of 80° longitude (except that of maize). Ragi as a kharif cereal (2.4 m ha) is mainly concentrated in Karnataka, Tamil Nadu and Andhra Pradesh which account for main than 60 per cent of the total area under this harvest in India. The harvesting patterns based on each of these kharif cereals are discussed.

The Maize-based Harvesting Patterns

The largest area under the kharif maize is in Uttar Pradesh (1.4m ha), followed by Bihar (0.96 m ha), Rajasthan (0.78 m ha), Madhya Pradesh (0.58 m ha) and Punjab (0.52 m ha). In four states namely Gujarat, Jammu and Kashmir, Himachal Pradesh and Andhra Pradesh, the area under maize ranges from 0.24 to 0.28 m ha in each, whereas other states have much less area under it. Taking the rainfall of the maize growing areas under consideration, over 72 per cent of the areas receive 20-30 cm per month of rainfall in at least two months or more during the south west monsoon season. On the all-India basis, about 12 harvesting patterns has been identified.

They have maize as the base harvest. In the maize growing areas of Uttar Pradesh and Bihar, rice in kharif and wheat in rabi are the main alternative harvests. In some areas, Bajra, groundnut, sugarcane, Ragi and pulses are taken as alternative harvests. In Rajasthan maize is grown as an extendedharvest in some areas, whereas in other places, it is replaced by small millets, pulses, groundnut and wheat (Rabi) as alternative harvest.

InMadhya Pradesh mainly the kharifgear is replaced by maize, whereas rice and groundnut are also grown to a limited extent. In Punjab maize has groundnut, fodder harvests and wheat (Rabi) as alternative harvests. In other states, *e.g.* Gujarat, rice, groundnut, cotton and wheat form the alternative harvests in the maize-growing areas of Himachal Pradesh, whereas in Andhra Pradesh, rice, kharif jowar, and oilseeds are grown in these areas.

The Kharifjowar-based Harvesting Patterns

The area under the kharifjowar in India is highest in Maharashtra (2.5 m ha), closely followed by Madhya Pradesh (2.3 m ha), whereas in each of the states of Rajasthan, Andhra Pradesh, Karnataka and Gujarat, the area under this harvest is between 1.0 and 1.4 m ha. Jowar is mainly grown where rainfall distribution ranges from 10-20 per month at least for 3 to 4 months of the south-westerly monsoon or is still more abundant.

On the all-India basis, about 17 major harvesting patterns has been identified. In them the base harvestsarekharif. Most of the alternative harvests are also of the type which can be grown under medium rainfall.

In Maharashtra cotton, pulses, groundnut and small millets are sown as alternative harvests. In the adjacent states of Madhya Pradesh, besides the above alternative harvests, wheat and fodder are sown. In Rajasthan wheat, cotton, Bajra and maize are grown in the kharif-Jowar tract, whereas in Andhra Pradesh, groundnuts, cotton, oilseeds and pulses form the main alternative harvests. Besides cotton and groundnut, ragi is sown in the kharif-jowar tarct of Karnataka, whereas in Gujarat, bajra, cotton and groundnut are the major alternative harvests.

The Bajra-based Harvesting Patterns

Bajra is more drought-resistantharvest than several other cereal harvests and is generally preferred in low-rainfall areas and on light soils. The area under the Bajraharvest in India is about 12.4 m ha and Rajasthan (4.6 m ha) shares about the 2/3 total area. Maharashtra, Gujarat and Uttar Pradesh together have over 4.6 m ha, constituting an additional 1/3 area under Bajra, in India. Over 66 percent of this harvest is grown in areas receiving 10-20 cm per month of rainfall, extending over 1 to 4 months of the southwesterly monsoon.

On the all-India basis, about 20 major harvesting patterns have been identified with Bajra. However, it may be observed that jowar and bajra are grown mostly under identical environmental conditions and both have a wide spectrum adaptability in respect of rainfall, temperature and rainfall.

Considering the harvesting patterns in different states, Bajra is grown along with pulses, groundnut, oilseeds and kharifgrower in Rajasthan. Gujarat has a similar harvesting pattern in its Bajra areas, except that cotton and tobacco are also grown. In Maharashtra besides having some areas solely under Bajra, pulses, wheat, Rabijowar, groundnut and cotton are substituted for it.

The Groundnut Based Harvesting Patterns

Groundnut is sown over an area of about 7.2 m ha, mostly in five major groundnut-producing states of Gujarat (24.4 per cent area), Andhra Pradesh (20.2) per cent), Tamil Nadu (13.5 per cent), Maharashtra (12.2 per cent) and Karnataka (12.0 per cent). Five other states *viz.* Madhya Pradesh, Uttar Pradesh, Punjab, Rajasthan and Orissa together have about 17.3 per cent of the total area under this harvest.

The rainfall in the groundnut area ranges from 20-30 cm per month in one of the monsoon months and much less in the other months. In some cases the rainfall is even less than 10 cm. Per month during the growth of the harvest. The irrigated area under groundnut is very small and that too, in a few states only, *viz.* Punjab(16.4 per cent), Tamil Nadu (13.3 per cent)and Andhra Pradesh (12.5 per cent).

On the all-India level, about 9 harvesting patterns has been identified with this harvest. In Gujarat besides the sole harvest of groundnut in some areas, Bajra, is the major alternative harvest, whereas the kharifjowar, cotton and pulses are also grown in this tract. In Andhra Pradesh and Tamil Nadu, this harvest receives irrigation in some areas and rice forms an alternative harvest. Under rain-fed conditions, Bajra, kharifjowar, small millets, cotton and pulses are grown as alternative harvests. In Maharashtra both the kharif and Rabi jowar and small millets are important alternative harvests. In Karnataka also, the grower is the major alternative harvest, whereas cotton, tobacco, sugarcane and wheat are also grown in this tract.

The Cotton-based Harvesting Patterns

Cotton is grown over 7.6 m ha in India. Maharashtra shares 36 percent (2.8 m ha), followed by Gujarat with 21 per cent (1.6 m ha), Karnataka with 13 per cent (1 m ha) and Madhya Pradesh with 9 per cent (0.6 m ha) of the area. Together, these four states account for about 80 per cent of the area under cotton. Other Cottom-growing states with smaller areas are Punjab, with 5 per cent (0.4 m ha), Andhra Pradesh and Tamil Nadu each with 4 per cent (0.31 m ha), Haryana and Rajasthan with 3 per cent of each (0.2 m ha each). Most of the cotton areas in the country are under the high to medium rainfall zone. The cotton grown in Madhya Pradesh, Maharashtra, Karnataka, and Andhra Pradesh (4.8 m ha) is rain-fed, whereas in Gujarat and Tamil Nadu (1.93 m ha) it receives partial irrigation 16-20 per cent of the area). The area under cotton in Punjab, Haryana, Rajasthan and Uttar Pradesh (0.8 m ha) gets adequate irrigation, ranging from 71 to 97 per cent of the area. These growing conditions, together with the species of cotton grown, determine the duration of the harvest which may vary from about 5 to 9 months.

On the all-India basis, about 16 broad harvesting pattens has been identified. In Maharashtra, Madhya Pradesh, Andhra Pradesh and Karnataka, the harvesting patterns in the cotton-growing areas are mostly similar owing to identical rainfall. These patterns include Jowar (kharif and Rabi), groundnut and small millets. Pulses and wheat are also grown in a limited area. In some pockets, where irrigation is available, rice and sugarcane are also grown. In Gujarat, rice, tobacco and maize are grown, besides the rain-fedharvests, *e.g.* Jowar and Bajra.

THE RABI SEASON HARVESTPING PATTERNS

Among the rabi harvests, wheat, together with barley and oats, jowar and gram, are the main base harvests among the rabi harvesting patterns. Generally, wheat and gram are concentrated in the subtropical region in northern India, whereas the Rabi sorghum is grown mostly in the Deccan. The extent of these areas in different states is as follows;

Harvest	Area	Region (per cent of all-India area)
Sugar cane	2.5 m ha	Uttar Pradesh (51), Haryana (6),Bihar (6), Punjab (6), Maharashtra (8), Andhra Pradesh (5),Tamil Nadu (5), Karnataka (3)
Tobacco	0.427 m ha	Andhra Pradesh (48), Gujarat (19.5), Karnataka (8.7), Maharashtra (3.5), Tamil Nadu (3.5)
Potato	0.491	Uttar Pradesh (33.6), Bihar (20.4), West Bengal (13.3), Assam (5.2), Orissa (4.8)
Jute	0.778	West Bengal (60), North East Region (18.7), Bihar (17.6), Orissa (6.1), Uttar Pradesh (1.7)
Coconut	1.05 m ha	Kerala (68.3), Karnataka (12.4), Tamil Nadu (9.7), Andhra Pradesh (3.5)
Rubber	0.197 m ha	Kerala (92.8), Tamil Nadu (5.0), Karnataka (1.9)
Cashew	0.264 m ha	Kerala (67.4), Karnataka (12.1), Andhra Pradesh (10.8), Tamil Nadu (9.8), Maharashtra (4.8)
Tea	0.35 m ha	West Bengal, Assam and Tripura (77), Kerala, Tamil Nadu and Karnataka (20)
Coffee	0.138 m ha	Kerala , Tamil Nadu and Karnataka (99)
All fruit-harvests"	1.8 m ha	Spread allover India
Onion	0.16 m ha	Maharashtra (18.5), Karnataka (11.7), Andhra Pradesh (12.8), Tamil Nadu (11.2), West Bengal (7.6), Madhya Pradesh (7.2), Orissa (6.8), Punjab (6.2)
Chillies	0.733 m ha	Andhra Pradesh (26.9), Maharashtra (20.4), Karnataka(14.5), Madhya Pradesh (5.5), Tamil Nadu (10.1)
Coriander	0.283 m ha	Andhra Pradesh (36), Rajasthan (23.6), Madhya Pradesh (11.1), Tamil Nadu (10.0)

In several sugarcane-growing areas, mono-harvesting is practiced, and during the interval between the harvests, short duration seasonal harvests are grown. In U.P., Bihar, Punjab and Haryana, wheat and maize are the rotation harvests. Rice is also grown in some areas. In the southern states, namely Tamil Nadu, Karnataka and Andhra Pradesh, Ragi, rice and pulses are grown along with sugar cane. In Maharashtra, pulses, jowar and cotton are grown. In the potato-growing region, maize, pulses, wheat is the alternative harvests. In the tobacco-growing areas, depending on the season and the type of tobacco, jowar, oilseeds and maize are grown in rotation. In the jute-growing areas, rice is the usual alternative harvest.

In the case of the plantation-harvests, interharvesting with pulses and fodder harvests is common.Spices and condiments are generally grown on fertile soils. Chillies are rotated with jowar, whereas onion, corriander, turmeric and ginger are grown as mixed harvests with other seasonal harvests.

MIXED HARVESTPING

Harvests mixtures are widely grown, especially during the kharif season. Pulses and some oilseeds are grown with maize, jowar and Bajra. Lowland rice is invariably grown unmixed, but in the case of upland rice, several mixtures are prevalent in eastern Uttar Pradesh, with the Chotanagpur Division of Bihar and in the Chhatisgarh Division of Madhya Pradesh. During the rabi season, especially in the unirrigated area of the north, wheat and barley and wheat and gram or wheat + barley + gram are the mixtures of grain harvests. Brassica and safflower are grown mixed with gram or even with wheat. Mixed harvesting was considered by researchers a primitive practice, but now many researchers regard mixed harvesting as the most efficient way of using land. Several new mixtures have recently been suggested. They ensure an efficient utilisation of sunshine and land. Breeders are developing plant types in pulses and oilseeds, with good compatibility with row harvests.

THE FUTURE OF HARVESTING PATTERNS

With the increase in population, the irrigated area is increasing and with advances in farm science, most of the extensive harvesting patterns are giving way to intensive harvesting. The development in minor irrigation works has especially provided the farmers with opportunities to harvest their land all the year round with high-yielding varieties. This intensive harvesting will require an easy and the ready availability of balanced fertilizers and plant protection chemicals and an appropriate price policy for inputs and farm produce.

India is a country of small farmers. In the future the size of the holdings will diminish further. The country has to produce enough for its people without deteriorating the quality of the environment. This is the challenge of the future for the farmers, farm scientists, extension workers and administrators.

LAND RELATIONS IN PRE-BRITISH INDIA

Indian farming began in 9000 BCE as a result of early cultivation of plants, and domestication of harvests and animals. Settled life soon followed with implements and techniques being developed for farming. Double monsoons led to two harvests being reaped in one year. Indian products soon reached the world via existing trading networks and foreign harvests were introduced to India. Plants and animals—considered essential to their survival by the Indians—came to be worshiped and venerated.

The middle ages saw irrigation channels reach a new level of sophistication in India and Indian harvests affecting the economies of other regions of the world under Islamic patronage. Land and water management methods were developed with an aim of providing uniform growth. Despite some stagnation during the later modern era the independent Republic of India was able to develop a comprehensive farmprogram.

EARLY HISTORY

Wheat, barley and jujube were domesticated in the Indian subcontinent by 9000 BP. Domestication of sheep and goat soon followed. This period also saw the first domestication of the elephant. Barley and wheat cultivation—along with the domestication of cattle, primarily sheep and goat—was visible in Mehrgarh by 8000-6000 BCE. Agro pastoralism in India included threshing, planting harvests in rows—either of two or of six—and storing grain in granaries. In the period of the Neo-lithic revolution (roughly 8000-5000 BCE.), Farming was far from the dominant mode of support for human societies. But those who adopted it, have survived and increased, and passed their techniques of production to the next generation. This transformation of knowledge was the base of further development in farming. By the 5th millennium BCE farm communities became widespread in Kashmir.

Zaheer Baber (1996) writes that 'the first evidence of cultivation of cotton had already developed'. Cotton was cultivated by the 5th millennium BCE-4th millennium BCE. The Indus cotton industry was well developed and some methods used in cotton spinning and fabrication continued to be practiced till the modern Industrialisation of India.

A variety of tropical fruit such as mango and muskmelon are native to the Indian subcontinent. The Indians also domesticated hemp, which they used for a number of applications including making narcotics, fibre, and oil. The farmers of the Indus Valley, which thrived in modern-day Pakistan and North India, grew peas, sesame, and dates. Sugar cane was originally from tropical South Asia and Southeast Asia. Different species likely originated in different locations with S. Barberioriginates in India and S. edule and S. officinarum coming from New Guinea.

Wild Oryza rice appeared in the Belan and Ganges valley regions of northern India as early as 4530 BCE and 5440 BCE respectively.

Rice was cultivated in the Indus Valley Civilization. Farm activity during the second millennium BC included rice cultivation in the Kashmir and Harrappan regions. Irrigation was developed in the Indus Valley Civilization by around 4500 BCE. The size and prosperity of the Indus civilization grew as a result of this innovation, which eventually led to more planned settlements making use of drains and sewers. Sophisticated irrigation and water storage methods were developed by the Indus Valley Civilization, including artificial reservoirs at Girnar dated to 3000 BCE, and an early canal irrigation method from circa 2600 BCE. Archeological evidence of an animal-drawn plough dates back to 2500 BC in the Indus Valley Civilization.

VEDIC PERIOD – POST MAHA JANAPADAS PERIOD (1500 BCE – 200 CE)

Gupta (2004) finds it likely that summer monsoons may have been longer and may have contained moisture in excess than required for normal food production. One effect of this excessive moisture would have been to aid the winter monsoon rainfall required for winter harvests. In India, both wheat and barley are held to be Rabi (winter) harvests and—like other parts of the world—would have largely depended on winter monsoons before the irrigation became widespread. The growth of the Kharifharvests would have probably suffered as a result of excessive moisture. Jute was first cultivated in India, where it was used to make ropes and cordage. Some animals—thought of the Indians as being vital to their survival—came to be worshiped. Trees were also domesticated, worshiped, and venerated—Pipal and Banyan in particular. Others came to be known for their medicinal uses and found mention in the holistic medical method Ayurveda.

EARLY COMMON ERA – HIGH MIDDLE AGES (200–1200 CE)

The Tamil people cultivated a wide range of harvests such as rice, sugarcane, millets, black pepper, various grains, coconuts, beans, cotton, plantain, tamarind and sandalwood. Jackfruit, coconut, palm, areca and plantain trees were also known. Method attic ploughing, manuring, weeding, irrigation and harvest protection were practiced for sustained farming. Water storage methods were designed during this period. Kallanai (1st-2nd century CE), a dam built on river Kaveri during this period, is considered the as one of the oldest water-regulation structures in the world still in use.

Spice trade involving spices native to India—including cinnamon and black pepper—gained momentum as India starts shipping spices in the Mediterranean. Roman trade with India followed as detailed by the archaeological record and the Periplus of the Erythraean Sea. Chinese sericulture attracted Indian sailors during the early centuries of the common era. Crystallised sugar was discovered by the time of the Guptas (320-550 CE), and the earliest reference of candied sugar come from India. The process was soon transmitted to China with traveling Buddhist monks. Chinese documents confirm at least two missions to India, initiated in 647 CE, for obtaining technology for sugar-refining. Each mission returned with the results on refining sugar. Indian spice exports find mention in the works of Ibn Khurdadhbeh (850), al-Ghafiqi (1150), Ishak bin Imaran (907) and Al Kalkashandi (fourteenth century). Noboru Karashima's research of the agrarian society in South India during the Chola Empire (875-1279) reveals that during the Chola rule land was transferred and collective holding of land by a group of people slowly gave way to individual plots of land, each with their own irrigation method. The growth of individual disposition of farming property may have led to a decrease in areas of dry cultivation. The Cholas also had bureaucrats which oversaw the distribution of water—-particularly the distribution of water by tank-and-channel networks to the drier areas.

LATE MIDDLE AGES – EARLY MODERN ERA (1200–1757 CE)

The construction of waterworks and aspects of water technology in India is described in Arabic and Persian works. The diffusion of Indian and Persian irrigation technologies gave rise to an irrigation method which brought about economic growth and growth of material culture. Farmers 'zones' were broadly divided into those producing rice, wheat or millets. Rice production continued to dominate Gujarat and wheat dominated north and central India. Land management was particularly strong during the regime of Akbar the Great (reign: 1556-1605), under whom scholar-bureaucrat Todarmal formulated and implemented elaborated methods for farm management on a rational basis. Indian harvests—such as cotton, sugar, and citric fruits—spread visibly throughout North Africa, Islamic Spain, and the Middle East. Though they may have been in cultivation prior to the solidification of Islam in India, their production was further improved as a result of this recent wave, which led to far-reaching economic outcomes for the regions involved.

2

British Agricultural Revolution

The British agricultural revolution describes a period of agricultural development in Britain between the 16th century and the mid-19th century, which saw a massive increase in agricultural productivity and net output. This in turn supported unprecedented population growth, freeing up a significant percentage of the workforce, and thereby helped drive the Industrial Revolution. How this came about is not entirely clear. In recent decades, historians cited four key changes in agricultural practices, enclosure, mechanization, four-field crop rotation, and selective breeding, and gave credit to a relatively few individuals.

CHALLENGES AND ISSUES

The challenges and issues of industrial agriculture for global and local society, for the industrial agriculture industry, for the individual industrial agriculture farm, and for animal rights include the costs and benefits of both current practices and proposed changes to those practices. Current industrial agriculture practices are temporarily increasing the carrying capacity of the Earth for humans while slowly destroying the *long term* carrying capacity of the earth for humans necessitating a shift to a sustainable agriculture form of industrial agriculture.

Fig. The British Agricultural Revolution describes a Period of Agricultural Development

This is a continuation of thousands of years of the invention and use of technologies in feeding ever growing populations. [W]hen hunter-gatherers with growing populations depleted the stocks of game and wild foods across the Near East, they were forced to introduce agriculture. But agriculture brought much longer hours of work and a less rich diet than hunter-gatherers enjoyed. Further population growth among shifting slash-and-burn farmers led to shorter fallow periods, falling yields and soil erosion. Plowing and fertilizers were introduced to deal with these problems-but once again involved longer hours of work and degradation of soil resources(Boserup, The Conditions of Agricultural Growth, Allen and Unwin, 1965, expanded and updated in Population and Technology, Blackwell, 1980.).

While the point of industrial agriculture is lower cost products to create greater productivity thus a higher standard of living as measured by available goods and services, industrial methods have side effects both good and bad. Further, industrial agriculture is not some single indivisible thing, but instead is composed of numerous separate elements, each of which can be modified, and in fact is modified in response to market conditions, government regulation, and scientific advances. So the question then becomes for each specific element that goes into an industrial agriculture method or technique or process: What bad side effects are bad enough that the financial gain and good side effects are outweighed? Different interest groups not only reach different conclusions on this, but also recommend differing solutions, which then become factors in changing both market conditions and government regulations.

Major Challenges and Issues Faced by Society

The major challenges and issues faced by society concerning industrial agriculture include:

Maximizing the benefits:

- Cheap and plentiful food
- Convenience for the consumer
- The contribution to our economy on many levels, from growers to harvesters to processors to sellers while minimizing the downsides:
- Environmental and social costs
- Damage to fisheries
- Cleanup of surface and groundwater polluted with animal waste
- Increased health risks from pesticides
- Increased ozone pollution and global warming from heavy use of fossil fuels

Benefits of Cheap and Plentiful Food

Population (est.) 10,000 BCE – 2000 CE. Very roughly:

- 30,000 years ago hunter-gatherer behaviour fed 6 million people

- 3,000 years ago primitive agriculture fed 60 million people
- 300 years ago intensive agriculture fed 600 million people
- Today industrial agriculture feeds 6000 million people

Table. Estimated world population at various dates, in thousands

Year	*World*	*Africa*	*Asia*	*Europe*	*Central & South America*	*North America*	*Oceania*
8000 BCE	8 000						
1000 BCE	50 000						
500 BCE	100 000						
1 CE	200,000 plus						
1000	310 000						
1750	791 000	106 000	502 000	163 000	16 000	2 000	2 000
1800	978 000	107 000	635 000	203 000	24 000	7 000	2 000
1850	1 262 000	111 000	809 000	276 000	38 000	26 000	2 000
1900	1 650 000	133 000	947 000	408 000	74 000	82 000	6 000
1950	2 518 629	221 214	1 398 488	547 403	167 097	171 616	12 812
1955	2 755 823	246 746	1 541 947	575 184	190 797	186 884	14 265
1960	2 981 659	277 398	1 674 336	601 401	209 303	204 152	15 888
1965	3 334 874	313 744	1 899 424	634 026	250 452	219 570	17 657
1970	3 692 492	357 283	2 143 118	655 855	284 856	231 937	19 443
1975	4 068 109	408 160	2 397 512	675 542	321 906	243 425	21 564
1980	4 434 682	469 618	2 632 335	692 431	361 401	256 068	22 828
1985	4 830 979	541 814	2 887 552	706 009	401 469	269 456	24 678
1990	5 263 593	622 443	3 167 807	721 582	441 525	283 549	26 687
1995	5 674 380	707 462	3 430 052	727 405	481 099	299 438	28 924
2000	6 070 581	795 671	3 679 737	727 986	520 229	315 915	31 043
2005	6 453 628	887 964	3 917 508	724 722	558 281	332 156	32 998**

An example of industrial agriculture providing cheap and plentiful food is the U.S.'s "most successful programme of agricultural development of any country in the world". Between 1930 and 2000 U.S. agricultural productivity (output divided by all inputs) rose by an average of about 2 percent annually causing food prices paid by consumers to decrease. "The percentage of U.S. disposable income spent on food prepared at home decreased, from 22 percent as late as 1950 to 7 percent by the end of the century."

Convenience and Choice

Industrial agriculture treats farmed products in terms of minimizing inputs and maximizing outputs at every stage from the natural resources of sun, land and water to the consumer which results in a vertically integrated industry that genetically manipulates crops and livestock; and processes, packages, and markets in whatever way generates maximum return on investment creating convenience foods many customers will pay a premium for. A consumer backlash against food sold for taste, convenience, and profit rather than nutrition and other values (*e.g.* reduce waste, be natural, be ethical) has led the industry to also provide organic food, minimally processed foods, and minimally packaged foods to maximally satisfy all segments of society thus generating maximum return on investment.

Industrial agriculture uses huge amounts of water, energy, and industrial chemicals; increasing pollution in the arable land, useable water and atmosphere. Herbicides, insecticides, fertilizers, and animal waste products are accumulating in ground and surface waters. "Many of the negative effects of industrial agriculture are remote from fields and farms. Nitrogen compounds from the Midwest, for example, travel down the Mississippi to degrade coastal fisheries in the Gulf of Mexico. But other adverse effects are showing up within agricultural production systems — for example, the rapidly developing resistance among pests is rendering our arsenal of herbicides and insecticides increasingly ineffective."

A study done for the US. Office of Technology Assessment conducted by the UC Davis Macrosocial Accounting Project concluded that industrial agriculture is associated with substantial deterioration of human living conditions in nearby rural communities.

"Confined animal feeding operations" or "intensive livestock operations" or "factory farms", can hold large numbers (some up to hundreds of thousands) of animals, often indoors. These animals are typically cows, hogs, turkeys, or chickens. The distinctive characteristics of such farms is the concentration of livestock in a given space. The aim of the operation is to produce as much meat, eggs, or milk at the lowest possible cost.

Food and water is supplied in place, and artificial methods are often employed to maintain animal health and improve production, such as therapeutic use of antimicrobial agents, vitamin supplements and growth hormones. Growth hormones are not used in chicken meat production nor are they used in the European Union for any animal. In meat production, methods are also sometimes employed to control undesirable behaviours often related to stresses of being confined in restricted areas with other animals. More docile breeds are sought (with natural dominant behaviours bred out for example), physical restraints to stop interaction, such as individual cages for chickens, or animals physically modified, such as the de-beaking of chickens to reduce the harm of fighting. Weight gain is encouraged by the provision of plentiful supplies of food to animals breed for weight gain.

The designation "confined animal feeding operation" in the U.S. resulted from that country's 1972 Federal Clean Water Act, which was enacted to protect and restore lakes and rivers to a "fishable, swimmable" quality. The United States Environmental Protection Agency (EPA) identified certain animal feeding operations, along with many other types of industry, as point source polluters of groundwater. These operations were designated as CAFOs and subject to special anti-pollution regulation. In 24 states in the U.S., isolated cases of groundwater contamination has been linked to CAFOs. For example, the ten million hogs in North Carolina generate 19 million tons of waste per year. The U.S. federal government acknowledges the waste disposal issue and requires that animal waste be stored in lagoons. These lagoons can be as large as 7.5 acres (30,000 m). Lagoons not protected with an impermeable liner can leak waste into groundwater under some conditions, as can runoff from manure spread back onto fields as fertilizer in the case of an unforeseen heavy rainfall. A lagoon that burst in 1995 released 25 million gallons of nitrous sludge in North Carolina's New River. The spill allegedly killed eight to ten million fish.

ADVANCES THAT HELPED THE AGRICULTURE REVOLUTION

Very few people will work harder or smarter if their work or income can arbitrarily be seized by those in power. One of the keys to the British Agricultural Revolution was the traditional restricted role of kings and the so-called "aristocrats" as exemplified by a long legal tradition going back to at least the signing of the Magna Carta by King John of England in 1217. The British Agricultural Revolution relied in large part on the establishment of a democratic government that used Parliament-passed laws, English Common law, the established Courts with Judicial independence and the rule of law to protect life, liberty and property in England, Wales, Scotland, etc. Empowering the farmers, investors, inventors and businessman was accomplished by increasingly and successfully restricting the power the Pope and the king and the so-called *"aristocracy"* had in England.

King Henry VIII of England separated the Church of England from the often repressive rule of the Pope and the Roman Catholic Church and its many bishops, clerics, Ecclesiastical courts, etc. in 1535. In the English Civil War (1640–1649) the Parliamentarians under Oliver Cromwell (1599–1658) defeated and executed the king, Charles I of England, and abolished about 700 king granted monopolies. After 1660 when King Charles II of England (1630–1685) was restored and then again after 1685 when King James II of England came to power there was an on-going battle between king and the English Parliament. This was resolved in the Parliament's favor in 1688 when the Catholic Stuart King James II of England was forced to flee for his life to France.

At the end the Glorious Revolution in 1688, the new king and queen, William and Mary agreed to abolish nearly all remaining monopolies and signed the English Bill of Rights 1689 and eliminated the hearth tax marking a new level co-operation and power sharing between the Parliament and the English monarchs. All of these processes led to a greater measure of legal protection for life, liberty, and property in England that encouraged and empowered the middle class at the beginning of the Agricultural and Industrial Revolutions. The Glorious Revolution of 1688 signaled the lessening of several centuries of tension and conflict between the crown and parliament, and the end of the idea that English kings had any *divine rights* and that England, etc. would be restored to Roman Catholicism. The new King William III and his wife Mary II were Protestant leaders from the Dutch Republic that were invited by Parliament to rule England. To make up for the loss of tax revenue, due to the cancellation of the hearth tax, uniform property taxes were imposed with few exclusions.

With legal assurances that their property and work would not be arbitrarily confiscated without legal proceedings by the king, the so-called "nobility", church or anybody else investors and inventors could secure capital to invest or use *sweat equity* to build improved farms or make other improvements. All realized some taxes were necessary to maintain law and order, pay for a government bureaucracy, tax collectors, monarchy, an Army and a Navy, post office system, and protect life, liberty and property by running an independent court system. Surprisingly, the government was run with an average tax rate of about 10% of GDP with taxes rising to about 20% of GDP in the Seven Years' War, American Revolutionary War and the Napoleonic Wars and then receding back to about 10% when the war bonds, loans etc. were paid off.

Private investors financed and built transportation systems like sailing ships, paddle steamers (after 1830), steam ships (after 1860), refrigerator ships (after 1890), toll roads (after 1700), canals (after 1760), railroads (after 1830) allowed people, goods, foods and animal crops to be gathered and shipped cheaply and with increasing rapidity over ever increasing distances. New food preservation techniques like canning (after 1800), refrigeration (after 1880) as well as traditional food storage techniques like root cellars, granaries, etc. were improved. These allowed more food to be stored over longer periods of time and minimized the impact of local shortages.

Ever increasing literacy, spread in large part by the Protestant Reformation and its pressure to read and study the Bible in English, was well spread in England and even broader in Scotland. Public pressure and public and private funding were used to try and teach more and more males and some females to read, write and cipher. The fraction of the populace educated continually increased with about 30% of all males at least semi literate by 1630, rising to 60% by 1730 and 65% by 1830 before reaching 90% in 1900. English women literate rates were about 10% by 1700, 35% by 1800 and increasing to 90+% by 1900. Cheap and wide spread "How-To" books made possible by Johannes Gutenberg's invention of the printing press (1438) and the numerous and wide spread low cost books and other literature that were made available. Increased literacy led to increased (uncensored) printing of books magazines and newspapers to satisfy a growing demand. Information on successful new agricultural techniques, scientific discoveries, inventions, prices and information of surpluses and shortages were spread faster and over much wider territories by the wide availability of cheap (compared to hand copied) books, magazines and newspapers.

New scientific discoveries found by Sir Isaac Newton, Charles Darwin and many others were increasingly added to the school curriculum and studied by an increasingly widening student body. New technologies like the telegraph (after 1830) and the telephone (after 1880) greatly improved the speed and flow of information needed to keep up with prices, shortages and surpluses to know which crops to buy, ship or grow. New crop rotation systems involving turnips and clover (plus others) made it less necessary to have so much land lie fallow. New crops had more yield per acre and new improved livestock that could be fed though the winters gave larger yields of meat. Larger drives of livestock from Ireland, Wales and Scotland increased the available meat supply available to be purchased in the ever increasing cities. New irrigation systems called water meadows allowed longer growing seasons on pasture land and increased yield from hay fields allowing more animals to be raised. Improved use and new fertilizers in addition to "manure" helped maintain and improve the arable land or reclaim formerly "waste" land. These improvements and new crops in turn supported unprecedented population growth, freeing up a significant percentage of the workforce, and thereby helped drive the Industrial Revolution.

The Bank of England (founded 1694) helped the farmers get and use the capital (raised from other investors) needed to buy farm improvements and herds of livestock. A good financial system was needed to put capital investments back into the farms and livestock operations. Since it was much more economical to use a few herders as reasonable to bring herds of livestock the long distances from Ireland, Wales and Scotland to England for fattening and/or slaughter accumulating the money to buy these herds was much facilitated by a workable banking system. The wealth (capital) earned by the workers in the Industrial Revolution in turn allowed these workers to buy and pay for more agricultural farm and livestock production. The Industrial Revolution also made many things much cheaper so the farmers could purchase more extensive wardrobes, metal tools, new machines, etc. with the same earnings. The much cheaper textile manufacturing brought about by the Industrial Revolution made available cheap washable cotton underwear that improved personal hygiene and slowed down the spread of gastrointestinal diseases like cholera etc.

There were large variations in the location and time different agricultural innovations were introduced in the many different agricultural zones in Britain and the rest of the world. Advances in science, engineering and elementary botany encouraged the progression of the Agricultural Revolution in Britain and elsewhere. More professional farm management, more capital investment, better agricultural education, improved fertilizations, new improved crops or higher yield, mechanization of farm work from oxen, horses, steam power and then gasoline or diesel power, four-field crop rotation, and selective breeding of livestock for larger size and other desirable characteristics have been highlighted as important links to the Agricultural revolution.

Increased understanding of the important chemicals need for proper plant nutrition allowed new fertilizers to be imported and made. Eventually the establishment of a chemical fertilizer industry made it possible to restore the chemicals lost in growing crops to maintain the needed enhanced food plant growth.

New agricultural equipment like cultivators, plows, threshers, mowers, combines, balers, etc. were invented and powered by oxen, horses, steam power, then gasoline or diesel as the farm machinery became mechanized and allowed the farmers to become ever more productive. The farm transportation system moved from pack animals to carts and wagons pulled by oxen, to mulesor horses to trucks (after about 1915).

As the Industrial Revolution picked up speed in the early 1800s the metal technologies, power sources, metal working skills, monetary systems, investors, capital, etc. became available for agricultural continuation and improvement. Early successes led to ever increasing innovative agricultural machinery and labor saving devices. The percentage of the population in agricultural work decreased from about 80% in the 1300s to less than 2% today (in the developed world) as the agricultural "revolution" continues.

CONCENTRATION OF ANIMALS, ANIMAL WASTE AND DEAD ANIMALS

The large concentration of animals, animal waste, and dead animals in a small space poses ethical issues. Animal rights and animal welfare activists have charged that intensive animal rearing is cruel to animals. As they become more common, so do concerns about air pollution and ground water contamination, and the effects on human health of the pollution and the use of antibiotics and growth hormones.

One particular problem with farms on which animals are intensively reared is the growth of antibiotic resistant bacteria. Because large numbers of animals are confined in a small space, any disease would spread quickly, and so antibiotics are used preventively. A small percentage of bacteria are not killed by the drugs, which may infect human beings if it becomes airborne.

According to the U.S. Centers for Disease Control and Prevention (CDC), farms on which animals are intensively reared can cause adverse health reactions in farm workers. Workers may develop acute and chronic lung disease, musculoskeletal injuries, and may catch infections that transmit from animals to human beings.

The CDC writes that chemical, bacterial, and viral compounds from animal waste may travel in the soil and water. Residents near such farms report nuisances such as unpleasant smells and flies, as well as adverse health effects.

The CDC has identified a number of pollutants associated with the discharge of animal waste into rivers and lakes, and into the air. The use of antibiotics may create antibiotic-resistant pathogens; parasites, bacteria, and viruses may be spread; ammonia, nitrogen, and phosphorus can reduce oxygen in surface waters and contaminate drinking water; pesticides and hormones may cause hormone-related changes in fish; animal feed and feathers may stunt the growth of desirable plants in surface waters and provide nutrients to disease-causing micro-organisms; trace elements such as arsenic and copper, which are harmful to human health, may contaminate surface waters.

PERENNIAL AGRICULTURAL SYSTEMS

Permaculture is an approach to designing human settlements and perennial agricultural systems that mimic the relationships found in the natural ecologies. It was first developed by Australians Bill Mollison and David Holmgren and their associates during the 1970s in a series of publications. The word *permaculture* is a portmanteau of *permanent agriculture*, as well as *permanent culture*. Permaculture design principles extend from the position that "The only ethical decision is to take responsibility for our own existence and that of our children". The intent was that, by rapidly training individuals in a core set of design principles, those individuals could design their own environments and build increasingly self-sufficient human settlements — ones that reduce society's reliance on industrial systems of production and distribution that Mollison identified as fundamentally and systematically destroying Earth's ecosystems.

While originating as an agro-ecological design theory, permaculture has developed a large international following of individuals who have received training through intensive two week long 'permaculture design courses'. This 'permaculture community' continues to expand on the original ideas, integrating a range of ideas of alternative culture, through a network of training, publications, permaculture gardens, and internet forums. In this way, permaculture has become both a design system and a loosely defined philosophy or lifestyle ethic.

The term *permanent agriculture* was coined by Franklin Hiram King in his classic book from 1911, *Farmers of Forty Centuries: Or Permanent Agriculture in China, Korea and Japan*. In this context, permanent agriculture is understood as agriculture that can be sustained indefinitely.

This definition was supported by Australian P. A. Yeomans who introduced an observation-based approach to land use in Australia in the 1940s, based partially on his understanding of geology. Yeomans introduced Keyline Design as a way of managing the supply and distribution of water of a site. Holmgren based his EcoVillage design on the keyline principle. The work of Howard T. Odum was also an early influence, especially for Holmgren. Odum's work focused on system ecology, in particular the Maximum power principle, which examines the energy of a system and how natural systems tend to maximise the energy embodied in a system. For example, the total calorific value of woodland is very high with its multitude of plants and animals. It is an efficient converter of sunlight into biomass. A wheat field, on the other hand, has much less total energy and often requires a large energy input in terms of fertilizer. Another early influence was the work of Esther Deans, who pioneered No-Dig Gardening methods. Other recent influences include the VAC system in Vietnam which is a government supported system to build Vegetable Aquaculture and Animal enClosures that cycle resources.

Fig. Posted in Edible Forest Gardens, Perennial Farming Systems

In the mid 1970s, Australians Bill Mollison and David Holmgren started to develop ideas that they hoped could be used to create stable agricultural systems. This was a result of their perception of a rapidly growing use of destructive industrial-agricultural methods. They saw that these methods were poisoning the land and water, reducing biodiversity, and removing billions of tons of soil from previously fertile landscapes. A design approach called "permaculture" was their response and was first made public with the publication of *Permaculture One* in 1978.

The term *permaculture* initially meant "permanent agriculture" but was quickly expanded to also stand for "permanent culture" as it was seen that social aspects were an integral part of a truly sustainable system. Mollison and Holmgren are widely considered to be the co-originators of the modern permaculture concept.

After the publication of *Permaculture One*, Mollison and Holmgren further refined and developed their ideas by designing hundreds of permaculture sites and organizing this information into more detailed books. Mollison lectured in over 80 countries and his two-week Design Course was taught to many hundreds of students. By the early 1980s, the concept had moved on from being predominantly about the design of agricultural systems towards being a more fully holistic design process for creating sustainable human habitats.

By the mid 1980s, many of the students had become successful practitioners and had themselves begun teaching the techniques they had learned. In a short period of time permaculture groups, projects, associations, and institutes were established in over one hundred countries. In 1991 a four-part Television documentary by ABC productions called 'The Global Gardener' showed permaculture applied to a range of worldwide situations, bringing the concept to a much broader public. Excerpts are available online through YouTube.

Permaculture has developed from its origins in Australia into an international 'movement'. English permaculture teacher Patrick Whitefield, author of *The Earth Care Manual* and *Permaculture in a Nutshell*, suggests that there are now two strands of permaculture: a) Original and b) Design permaculture.

Original permaculture attempts to closely replicate nature by developing edible ecosystems which closely resemble their wild counterparts. Design permaculture takes the working connections at use in an ecosystem and uses them as its basis. The end result may not look as "natural" as a forest garden, but still has an underlying design based on ecological principles. Through close observation of natural energies and flow patterns efficient design systems can be developed. This has become known as Natural Systems Design.

ELEMENTS OF DESIGN

Permaculture principles draw heavily on the practical application of ecological theory to analyze the characteristics and potential relationships between design elements. Each element of a design is carefully analyzed in terms of its needs, outputs, and properties.

For example a chicken needs water, moderated microclimate, food and other chickens, and produces meat, eggs, feathers and manure and can help break the soil. Design elements are then assembled in relation to one another so that the products of one element feed the needs of adjacent elements. Synergy between design elements is achieved while minimizing waste and the demand for human labour or energy. Exemplary permaculture designs evolve over time, and can become extremely complex mosaics of conventional and inventive cultural systems that produce a high density of food and materials with minimal input.

While techniques and cultural systems are freely borrowed from organic agriculture, sustainable forestry, horticulture, agroforestry, and the land management systems of indigenous peoples, permaculture's fundamental contribution to the field of ecological design is the development of a concise set of broadly applicable organizing principles that can be transferred through a brief intensive training.

Modern Permaculture

Modern permaculture is a system design tool. It is a way of:

1. looking at a whole system or problem;
2. observing how the parts relate;
3. planning to mend sick systems by applying ideas learnt from long-term sustainable working systems;
4. seeing connections between key parts.

In permaculture, practitioners learn from the working systems of nature to plan to fix the damaged landscapes of human agricultural and city systems. This thinking applies to the design of a kitchen tool as easily to the re-design of a farm. Permaculture practitioners apply it to everything deemed necessary to build a sustainable future. Commonly, "Initiatives... tend to evolve from strategies that focus on efficiency (for example, more accurate and controlled uses of inputs and minimisation of waste) to substitution (for example, from more to less disruptive interventions, such as from biocides to more specific biological controls and other more benign alternatives) to redesign (fundamental changes in the design and management of the operation)." "Permaculture is about helping people make redesign choices: setting new goals and a shift in thinking that affects not only their home but their actions in the workplace, borrowings and investments". Examples include the design and employment of complex transport solutions, optimum use of natural resources such as sunlight, and "radical design of information-rich, multi-storey polyculture systems".

"This progression generally involves a shift in the nature of one's dependence — from relying primarily on universal, purchased, imported, technology-based interventions to more specific locally available knowledge and skill-based ones. This usually eventually also involves fundamental shifts in world-views, senses of meaning, and associated lifestyles." "My experience is that although efficiency and substitution initiatives can make significant contributions to sustainability over the short term, much greater longer-term improvements can only be achieved by redesign strategies; and, furthermore, that steps need to be taken at the outset to ensure that efficiency and substitution strategies can serve as stepping stones and not barriers to redesign..." (Hill 2000)

Design Innovation

The core of permaculture has always been in supplying a design toolkit for human habitation. This toolkit helps the designer to model a final design based on an observation of how ecosystems interact. A simple example of this is how the Sun interacts with a plant by providing it with energy to grow. This plant may then be pollinated by bees or eaten by deer. These may disperse seed to allow other plants to grow into tall trees and provide shelter to these creatures from the wind. The bees may provide food for birds and the trees provide roosting for them. The tree's leaves fall and rot, providing food for small insects and fungus. Such a web of intricate connections allows a diverse population of plant life and animals to survive by giving them food and shelter. One of the innovations of permaculture design was to appreciate the efficiency and productivity of natural ecosystems, to use natural energies (wind, gravity, solar, fire, wave and more) and seek to apply this to the way human needs for food and shelter are met. One of the most notable proponents of this design system has been David Holmgren, who based much of his permaculture innovation on zone analysis.

Core Values

Permaculture is a broad-based and holistic approach that has many applications to all aspects of life. At the heart of permaculture design and practice is a fundamental set of 'core values' or ethics which remain constant whatever a person's situation, whether they are creating systems for town planning or trade; whether the land they care for is only a windowbox or an entire forest. These 'ethics' are often summarised as;

- Earthcare – recognising that Earth is the source of all life, that Earth is our valuable home, and that we are a part of Earth, not apart from it.
- Peoplecare – supporting and helping each other to change to ways of living that do not harm ourselves or the planet, and to develop healthy societies.
- Fairshare (or placing limits on consumption)-ensuring that Earth's limited resources are used in ways that are equitable and wise.

Modern thought about permaculture began with the issue of sustainable food production. It started with the belief that for people to feed themselves sustainably, they need to move away from reliance on industrialised agriculture. Where industrial farms use technology powered by fossil fuels (such as gasoline, diesel and natural gas), and each farm specialises in producing high yields of a single crop, permaculture stresses the value of low inputs and diverse crops. The model for this was an abundance of small-scale market and home gardens for food production, and a main issue was food miles.

O'BREDIM Design Methodology

O'BREDIM is a mnemonic and acronym for *observation, boundaries, resources, evaluation, design, implementation and maintenance*.

- Observation allows you first to see how the site functions within itself, to gain an understanding of its initial relationships. Some people recommend a year-long observation of a site before anything is planted. During this period all factors, such as lay of the land, natural flora and so forth, can be brought into the design. A year allows the site to be observed through all seasons, although it must be realised that, particularly in temperate climates, there can be substantial variations between years.

- Boundaries refer to physical ones as well as to those your neighbours might place on you, for example.
- Resources include the people involved, funding, as well as what you can grow or produce in the future.
- Evaluation of the first three will then allow you to prepare for the next three. This is a careful phase of taking stock of what you have at hand to work with.
- Design is a creative and intensive process, and you must stretch your ability to see possible future synergetic relationships.
- Implementation is literally the ground-breaking part of the process when you carefully dig and shape the site.
- Maintenance is then required to keep your site at a healthy optimum, making minor adjustments as necessary. Good design will preclude the need for any major adjustment.

Patterns

The use of patterns both in nature and reusable patterns from other sites is often key to permaculture design. This echoes the Pattern language of Christopher Alexander used in architecture which has been an inspiration for many permaculture designers. All things, even the wind, the waves and the earth on its axis, moving around the Sun, form patterns. In pattern application, permaculture designers are encouraged to develop:

1. Awareness of the patterns that exist in nature (and how these function)
2. Application of pattern on sites in order to satisfy specific design needs.

"The application of pattern on a design site involves the designer recognising the shape and potential to fit these patterns or combinations of patterns comfortably onto the landscape" Sampson-Kelly. Branching can be used for the direction of paths, rather than straight paths with square angles. Lobe-like paths of the main path (known as keyhole paths) can be used to minimise waste and compaction of the soil. Permaculture zones are a way of organising design elements in a human environment based on the frequency of human use and plant or animal needs. Frequently manipulated or harvested elements of the design are located close to the house in zones one and two such as herbs for the kitchen, whereas chickens like to be close but need to be kept at a safe distance to reduce noise and contamination (unless they are house trained). Less frequently used or manipulated elements, and elements that benefit from isolation (such as wild species) are farther away.

Links and Connections

Also key to the permacultural design model is that useful connections are made between components in the final design. The formal analogy for this is a natural mature ecosystem. So, in much the same way as there are useful connections between Sun, plants, insects and soil there will be useful connections between different plants and their relationship to the landscape and humans.

Another innovation of the permaculture design is to design a landuse or other system that has multiple outputs. In terms of Holmgren's application of H.T. Odum's work, a useful connection is viewed as one that maximises power: that is, maximizes the rate of useful energy transformation. A comparison which illustrates this is between a wheat field and a forest. "It is not the number of diverse things in a design that leads to stability, it is the number of beneficial connections between these components" Mollison 1988.

In permaculture and forest gardening, seven layers are identified:

- The canopy
- Low tree layer (dwarf fruit trees)
- Shrubs
- Herbaceous
- Rhizosphere (root crops)
- Soil Surface (cover crops)
- Vertical Layer (climbers, vines)

An eighth layer, Mycosphere (fungi), is often included.

A mature ecosystem such as ancient woodland has a huge number of relationships between its component parts: trees, understory, ground cover, soil, fungi, insects and other animals. Plants grow at different heights. This allows a diverse community of life to grow in a relatively small space. Plants come into leaf and fruit at different times of year.

Floor in the time before the top canopy re-appears with the spring. A wood suffers very little soil erosion, as there are always roots in the soil. It offers a habitat to a wide variety of animal life, which the plants rely on for pollination and seed distribution. The productivity of such a forest, in terms of how much new growth it produces, exceeds that of the most productive wheat field. It is in this observation-of how much more productive a wood may be on far less fertilizer input-that the potential productivity of a permaculture design is modelled. The many connections in a wood contribute together to a proliferation of opportunities for amplifier feedbacks to evolve that in turn maximise energy flow through the system.

Here is a photo of a layered warm temperate garden in NSW, Australia (courtesy of PermacualtureVisions). There are several layers: the canopy layer is Inga Edulis (ice cream bean), the middle stratum contains plum and peach, mango, mulberry and nurse plants such as native wattle. There are shrubs such as sage and woody herbs, ground covers such as sweet potato and vines such as passion fruit and kiwi fruit. The tubers consist of onions and taro.

Polyculture is agriculture using multiple crops in the same space, in imitation of the diversity of natural ecosystems, and avoiding large stands of single crops, or monoculture. It includes crop rotation, multi-cropping, and inter-cropping. Alley cropping is a simplification of the layered system which typically uses just two layers, with alternate rows of trees and smaller plants. Permaculture Guilds are groups of plants, animals and microbacteria which work particularly well together.

These can be those observed in nature such as the White Oak guild which centers on the White Oak tree and includes 10 other plants. Native communities can be adapted by substitution of plants more suitable for human use. The Three Sisters of maize, squash and beans is a well known guild. The British National Vegetation Classification provides a comprehensive list of plant communities in the UK. Guilds can be thought of as an extension of companion planting.

Increase Edge

Permaculturists maintain that where vastly differing systems meet, there is an intense area of productivity and useful connections. The greatest example of this is the coast. Where the land and the sea meet there is a particularly rich area that meets a disproportionate percentage of human and animal needs. This is evidenced by the fact that the overwhelming majority of humankind lives within 100 km of the sea. So this idea is played out in permacultural designs by using spirals in the herb garden or creating ponds that have wavy undulating shorelines rather than a simple circle or oval. Edges between woodland and open areas have been claimed to be the most productive.

Perennial plants are often used in permaculture design. As they do not need to be planted every year they require less maintenance and fertilisers. They are especially important in the outer zones and in layered systems. Ken Fern of Plants For A Future has spent many years investigating suitable perennial plants. As has Wes Jackson of The Land Institute.

Many permaculture designs involve animals other than humans. Chickens can be used as a method of weed control and also as a producer of eggs, meat and fertilizer. Some types of agroforestry systems combine trees with grazing animals.

Some projects are critical of the use of animals. However not all permaculture sites farm the animals. The animals are pets and can be treated as co-habitators and co-workers of the site, eating foods normally unpalatable to people such as slugs, termites, being an integral part of the pest management by eating some pests, supplying fertilizer through their droppings and controlling some weed species.

Annual Monoculture (anti-pattern)

Annual monoculture such as a wheatfield can be considered a pattern to be avoided in terms of space (height is uniform) and time (crops grow at the same rate until harvesting). During growth and especially after harvesting the system is prone to soil erosion from rain. The field requires a hefty input of fertilizers for growth and machinery for harvesting. The work is more likely to be repetitive, mechanised and rely on fossil fuels.

No pattern should be hard and fast and depending on the design considerations they can be broken. An example of this is broadscale permaculture practiced at Ragmans Lane Farm, which has a component of annual farming. Here the amount of human involvement is a key factor influencing the design.

Applying these values means using fewer non-renewable sources of energy, particularly petroleum based forms of energy. Burning fossil fuels contributes to greenhouse gases and global warming; however, using less energy is more than just combatting global warming. Food production should be a fully renewable system; but using current agricultural systems this is not the case. Industrial agriculture requires large amounts of petroleum, both to run the equipment, and to supply pesticides and fertilizers. Permaculture is in part an attempt to create a renewable system of food production that relies upon minimal amounts of energy.

For example permaculture focuses on maximizing the use of trees (agroforestry) and perennial food crops because they make a more efficient and long term use of energy than traditional seasonal crops. A farmer does not have to exert energy every year replanting them, and this frees up that energy to be used somewhere else.

Traditional pre-industrial agriculture was labor intensive, industrial agriculture is fossil fuel intensive and permaculture is design and information intensive and petrofree. Partially permaculture is an attempt to work smarter, not harder; and when possible the energy used should come from renewable sources such as wind power, passive solar designs or biofuels.

A good example of this kind of efficient design is the chicken greenhouse. By attaching the chicken coop to a greenhouse you can reduce the need to heat the greenhouse by fossil fuels, as the chicken's bodies heat the area. The chickens scratching and pecking can be put to good use to clear new land for crops. Their manure can be used to fertilise the soil. Feathers could be used in compost or as a mulch. In a conventional factory situation all these chicken outputs are seen as a waste problem. So in factories cooled by huge air conditioners, the chicken waste is extracted. All the energy is focused on egg production. Thus it is a further principle of permaculture that "pollution is energy in the wrong place".

RESTATEMENTS OF THE PRINCIPLES OF PERMACULTURE APPEAR

These restatements of the principles of permaculture appear in David Holmgren's *Permaculture: Principles and Pathways Beyond Sustainability*;

- Observe and interact-By taking the time to engage with nature we can design solutions that suit our particular situation.
- Catch and store energy-By developing systems that collect resources when they are abundant, we can use them in times of need.
- Obtain a yield-Ensure that you are getting truly useful rewards as part of the work that you are doing.
- Apply self-regulation and accept feedback-We need to discourage inappropriate activity to ensure that systems can continue to function well.
- Use and value renewable resources and services-Make the best use of nature's abundance to reduce our consumptive behaviour and dependence on non-renewable resources.
- Produce no waste-By valuing and making use of all the resources that are available to us, nothing goes to waste.
- Design from patterns to details-By stepping back, we can observe patterns in nature and society. These can form the backbone of our designs, with the details filled in as we go.
- Integrate rather than segregate-By putting the right things in the right place, relationships develop between those things and they work together to support each other.
- Use small and slow solutions-Small and slow systems are easier to maintain than big ones, making better use of local resources and producing more sustainable outcomes.
- Use and value diversity-Diversity reduces vulnerability to a variety of threats and takes advantage of the unique nature of the environment in which it resides.
- Use edges and value the marginal-The interface between things is where the most interesting events take place. These are often the most valuable, diverse and productive elements in the system.
- Creatively use and respond to change-We can have a positive impact on inevitable change by carefully observing, and then intervening at the right time.

A basic principle is thus to "add value" to existing crops. A permaculture design therefore seeks to provide a wide range of solutions by including its main ethics as an integral part of the final value-added design. Crucially, it seeks to address problems that include the economic question of how to either make money from growing crops or exchange crops for labour such as in the LETS scheme. Each final design therefore should include economic considerations as well as give equal weight to maintaining ecological balance, making sure that the needs of people working on the project are met and that no one is exploited.

Community economics requires a balance between the three aspects that comprise a community: justice, environment and economics, also called the "triple bottom line", or "ecological-economics-ethics" (EEE) or "triple E". A cooperative farmer's market could be an example of this structure. The farmers are the workers and owners. Additionally, all economics are limited by their ecology. No economic system stands apart independently from its eco-system; therefore, all external costs must be considered when discussing economics. One way of doing this is through designing a system that has "multiple outputs".

For example, a wheat field interspersed with walnuts will reduce soil erosion, act as a windbreak and provide a walnut crop as well as a wheat crop. Managing two crops will be more interesting work. Here the system comes into conflict with conventional agriculture and economics. Interplanting trees in a wheat field reduces the wheat yield and makes the field harder to harvest using machinery, as the operator has to drive around the trees. Most farms specialise in a few crops at a time and seek to maximise surplus in order to increase profit. This surplus can only be maintained with a massive injection of fossil fuels.

John Robin has been one the strongest critics of permaculture, criticising it for its potential to spread environmental weeds. This reflects a divide between native plant advocates and permaculture. Permaculture in the tropics, as expressed in 'Permaculture: A Designer's Manual', did not produce significant amounts of food or fruit when applied in Northern New South Wales and Queensland, largely since the closed canopy is not conducive to fruit production. The system, whilst very healthy in and of itself, yielded very little produce.

Bill Mollison himself has criticised *itinerant teachers* of permaculture, who go on to teach after only a short course. At one point, Mollison unsuccessfully tried to trademark the term *permaculture* to prevent this practice.

Perhaps the strongest criticism of permaculture is found in the Review of Toby Hemenway's book *Gaia's Garden*, published in the Winter 2001 edition of the *Whole Earth Review*. In it, Greg Williams critiques the view that woods were more productive than farmland, based on the theory of ecological succession which says that net productivity declines as ecosystems mature.

He also criticised the lack of scientifically respectable data and questions whether permaculture is applicable to more than a small number of dedicated people. But Hemenway's response in the same magazine disputes Williams's claim on productivity as focusing on climax rather than on maturing forests, citing data from ecologist Robert Whittaker's book *Communities and Ecosystems*. Hemenway is also critical of Williams's characterisation of permaculture as simply forest gardening.

In the years since its conception, permaculture has become a successful approach to designing sustainable systems. Its adaptability and emphasis on meeting human needs means that it can be utilized in every climatic and cultural zone. However, at the moment the large proportion of practitioners are only likely to be inspired individuals and there is a distinct lack of broadscale permaculture projects. Nevertheless, permaculture has also been used successfully as a development tool to help meet the needs of indigenous communities facing degraded standards of living from exposure to free-market economics.

Africa

Zimbabwe has 60 schools designed using permaculture, with a national team working within the schools' curriculum development unit. The UN High Commissioner for Refugees (UNHCR) has produced a report on using permaculture in refugee situations after successful use in camps in Southern Africa and Republic of Macedonia. The Biofarming approach applied in Ethiopia has very similar features and can be considered permaculture. It is mainly promoted by the non-governmental organisation BEA, based in Addis Ababa.

OCEANIA

Australia

The development of permaculture co-founder David Holmgren's home plot at Melliodora, Central Victoria, has been well documented at his website and published in e-book format. Designed from permaculture principles, Crystal Waters is a socially and environmentally responsible, economically viable rural subdivision north of Brisbane, Australia. Crystal Waters was designed by Max Lindegger, Robert Tap, Barry Goodman and Geoff Young, and established in 1987.

It received the 1996 World Habitat Award (assessed by Dr Wally N'Dow) for its "pioneering work in demonstrating new ways of low impact, sustainable living". Eighty-three freehold residential and two commercial lots occupy 20% of the 259ha (640 acre) property. The remaining 80% is the best land, and is owned in common. It can be licensed for sustainable agriculture, forestry, recreation and habitat projects.

Tikopia

Tikopians practice an intensive permaculture system, similar in principle to forest gardening and the gardens of the New Guinea highlands. Their agricultural practices are strongly and consciously tied to the population density. For example, around 1600 AD, the people agreed to slaughter all pigs on the island and substitute fishing, because the pigs were taking too much food that could be eaten by people.

New Zealand

There are many well established living examples of permaculture practice in New Zealand. Rainbow Valley Farm is the premier model. Rainbow Valley Farm was established in 1988 by Joe Polaischer and Trish Allen. The 21 ha. organic farm was designed on permaculture principles and ethics.

ASIA

Indonesia

The Indonesian Development of Education and Permaculture assisted in disaster relief in Aceh, Indonesia after the 2004 Tsunami. They have also developed Wastewater Gardens, a small-scale sewage treatment systems similar to Reedbeds.

Thailand

The Panya Project, located in Mae Taeng, Chiang Mai, Thailand, is a sustainable living project implementing permaculture principals and hosting workshops in English and Thai. In fall 2006, the project hosted a PDC taught by Geoff Lawton of the Permaculture Research Institute of Australia, and subsequently installed over 500 meters of swales and a 2 million litre dam.

The Panya Project used permaculture to help regenerate what used to be a monocrop mango plantation, transforming it into what is called a "biodiverse food forest, organic farm and education centre". The Panya Project also incorporates what they call "natural building" into their design, *e.g.*, wattle, cob and adobe brick.

EUROPE

Cyprus

Two acres of land at Ayia Skepi Therapeutic Centre in Filani village, a drug rehabilitation centre about 25 km from Nicosia, are being developed by Emily Markides, Julia Yelton, Charles Yelton and the residents of the detoxification centre.

France

- Annual events
  - Introduction to Permaculture events
  - Permaculture design courses in French and English
  - Regular 'entre-aide' work days.
- Francophone Permaculture
- Francophone Permaculture Forum
- L'Université Populaire de Permaculture
- The French Permaculture Association.

United Kingdom

There are a number of example permaculture projects in the UK, including:

- Agroforestry Research Trust, a not-for-profit organisation based in Dartington, Devon that runs a 2-acre (8,100 m^2) forest garden and publishes the journal Agroforestry News
- Chickenshack Housing co-op, a fully mutual housing co-op established in 1995 using permaculture design principles. Based in rural North Wales, the community has 4 dwellings and 6 residents on a 5-acre (20,000 m^2) site. Features include a biomass and solar district heating scheme, a half-acre forest garden and various wildlife conservation and habitat creation strategies. The community is very active in regional sustainability projects such as the Machynlleth Transition Towns initiative. It runs occasional courses in permaculture design and regularly receives visits from interested parties.
- Middlewood Trust, a permculture-based farm in North Lancashire running courses in permaculture, crafts, forestry and sustainability
- Plants for a Future, a vegan-organic project based at Lostwithiel in Cornwall that is researching and trialing edible and otherwise useful plant crops for sustainable cultivation. Their online database features over 7,000 such species that can be grown within the UK. A collaborative version of the database is in development by the permaculture.info project.
- *Prickly Nut Woods*, a 10-acre (40,000 m^2) woodland near Haslemere, Surrey that is managed by Ben Law. He uses a 'whole system' permacultural approach, using a wide variety of woodland products and documenting a complex web of relationships. He built a house almost entirely using products from the woodland, which was featured in Channel 4's Grand Designs TV series. The project has a second, larger property in North Devon, for which it is seeking a new group to take over.
- Ragmans Lane, a 60-acre (240,000 m^2) farm in the Forest of Dean in Gloucestershire.
- The RISC Roof Garden, on top of a development education centre in Reading city centre and inspired by Robert Hart's permaculture forest garden in Shropshire, is an excellent example of urban permaculture design.. It is used by schools, educators and designers as an educational resource for sustainable development and is a member of the National Gardens Scheme. The garden is composed of dense plantings of over 180 species of edible and medicinal plants and is fed by rainwater and composted waste from the centre.
- Tir Penrhos Isaf, near Dolgellau, developed by Chris and Lyn Dixon since 1986.

Other projects tend to be more community oriented, particularly in urban areas. These include Naturewise, a north London based group that tends a number of forest gardens and allotments as well as running regular permaculture introductory and design courses; and Organiclea, a workers cooperative that is involved in developing local food-growing and distribution initiatives around the Walthamstow area of east London.

The Transition Towns movement initiated in Totnes and Kinsale by Rob Hopkins is underpinned by permaculture design principles in its attempts to visualise sustainable communities beyond peak oil. The UK Permaculture Association publishes an extensive directory of other projects and example sites throughout the country.

LATIN AMERICA

Nicaragua

Project Bona Fide is a 43-acre (170,000 m^2) site on the twin volcano island of Ometepe, Nicaragua. Project Bona Fide has been in development for nearly a decade, and has become an important centre for education and community development.

Infrastructural systems contain: natural buildings built with local materials, terraced and medicinal plant gardens, an extensive nursery, seed bank, developing fruit and nut orchards, food forests, native timber forestry, timber bamboo plantings, water-catchment, drip irrigation and ferrocement technologies, renewable energy systems, and composting toilets.

Outreach efforts include social programmes that provide educational opportunities based in ecological agriculture, community reforestation efforts that are supported by our seed bank and nursery, local seed and plant exchanges, a children's nutritional kitchen and an upcoming community centre.

Brazil

IPEC-Ecocentro at the Instituto de Permacultura e Ecovilas do Cerrado-the Institute of Permaculture and Ecovillage of the Cerrado

IPCP-Instituto de Permacultura Cerrado-Pantanal (Permaculture Institute of Cerrado-Pantanal), Campo Grande, MS. Specializing in interactive teaching of Permaculture and direct work with Indigenous communities within the Cerrado biome.

NORTH AMERICA

Canada

Kootenay Permaculture Institute, British Columbia http://www3.telus.net/permaculture

United States of America

- Permaculture Research Institute Minnesota
- In Santa Fe, New Mexico, the Permaculture Institute utilizes a hands on approach to education on topics such as landscape and building design as well as water systems.
- Promoting urban permaculture in Los Angeles is Path to Freedom
- The Edible Plant Project implements and promotes elements of permaculture through a nonprofit nursery and workshops in Gainesville, FL (home of University of Florida).
- The Northeastern Permaculture Network brings together enthusiasts in the northeastern United States and eastern Canada.
- The Seattle Permaculture Guild is active in that city.
- The Portland Permaculture Guild (PPG) is very active in Portland, Oregon. There are many PC gardens in and around the Portland area. There are many fine teachers in Portland and in Oregon, including teachers OF PC teachers. Also, Toby Hemenway, noted Permaculture author and teacher, lives in the Portland area.
- The Regenerative Design Institute (RDI) is a non-profit educational organization in Bolinas, California.
- Burlington Permaculture is an ad-hoc community group based out of Burlington Vermont. BP unites neighbors to promote urban agriculture and reforestation, enhance neighborhoods, and strengthen the web of community resources as we look beyond sustainability towards a vibrant healthy relationship with our landscape.
- The Round Mountain Institute in the Gunnison valley of Colorado is a nonprofit that is dedicated to sustainable agriculture high in the rocky mountains near the continental divide.
- The Urban Permaculture Guild implements and promotes elements of permaculture through educational workshops and projects in East Bay and San Francisco, CA.
- The Urban-Suburban Sustainability Initiative based in Belleville, Illinois is a local grassroots organization in the process of starting up. It's focus will be on permaculture, bioremedification, environmental education and Local Exchange Trading Systems.
- Indigenous Permaculture (IPP) revitalizes the relationship of communities to the earth, and operates as a collaborative of communities sharing information, resources, and tools.

Cuba

Cuba has in the past 18 years transformed its food production using low-input, or organic agriculture and, to some degree, permaculture. Havana produces up to 50% of its food requirements within the city limits, all of it is organic and produced by people in their homes, gardens and in municipal spaces.

TECHNOLOGICAL DEVELOPMENT OF THE GREEN REVOLUTION

The projects within the Green Revolution spread technologies that had already existed, but had not been widely used outside of industrialized nations. These technologies included pesticides, irrigation projects, and synthetic nitrogen fertilizer.

The novel technological development of the Green Revolution was the production of what some referred to as "miracle seeds." Scientists created strains of maize, wheat, and rice that are generally referred to as HYVs or "high-yielding varieties." HYVs have an increased nitrogen-absorbing potential compared to other varieties. Since cereals that absorbed extra nitrogen would typically lodge, or fall over before harvest, semi-dwarfing genes were bred into their genomes. Norin 10 wheat, a variety developed by Orville Vogel from Japanese dwarf wheat varieties, was instrumental in developing Green Revolution wheat cultivars. IR8, the first widely implemented HYV rice to be developed by IRRI, was created through a cross between an Indonesian variety named "Peta" and a Chinese variety named "Dee Geo Woo Gen."

With the availability of molecular genetics in Arabidopsis and rice the mutant genes responsible *(reduced height(rht), gibberellin insensitive (gai1)* and *slender rice (slr1))* have been cloned and identified as cellular signalling components of gibberellic acid, a phytohormone involved in regulating stem growth via its effect on cell division. Stem growth in the mutant background is significantly reduced leading to the dwarf phenotype.

Photosynthetic investment in the stem is reduced dramatically as the shorter plants are inherently more stable mechanically. Assimilates become redirected to grain production, amplifying in particular the effect of chemical fertilisers on commercial yield.

HYVs significantly outperform traditional varieties in the presence of adequate irrigation, pesticides, and fertilizers. In the absence of these inputs, traditional varieties may outperform HYVs. One criticism of HYVs is that they were developed as F1 hybrids, meaning they need to be purchased by a farmer every season rather than saved from previous seasons, thus increasing a farmer's cost of production.

IDEA AND PRACTICE OF SUSTAINABLE AGRICULTURE

The idea and practice of sustainable agriculture has arisen in response to the problems of industrial agriculture. Sustainable agriculture integrates three main goals: environmental stewardship, farm profitability, and prosperous farming communities. These goals have been defined by a variety of disciplines and may be looked at from the vantage point of the farmer or the consumer.

METHODS OF ORGANIC FARMING

Organic farming methods combine some aspects of scientific knowledge and highly limited modern technology with traditional farming practices; accepting some of the methods of industrial agriculture while rejecting others. Organic methods rely on naturally occurring biological processes, which often take place over extended periods of time, and a holistic approach; while chemical-based farming focuses on immediate, isolated effects and reductionist strategies.

Integrated Multi-Trophic Aquaculture is an example of this holistic approach. Integrated Multi-Trophic Aquaculture (IMTA) is a practice in which the by-products (wastes) from one species are recycled to become inputs (fertilizers, food) for another. Fed aquaculture (*e.g.* fish, shrimp) is combined with inorganic extractive (*e.g.* seaweed) and organic extractive (*e.g.* shellfish) aquaculture to create balanced systems for environmental sustainability (biomitigation), economic stability (product diversification and risk reduction) and social acceptability (better management practices).

ORGANIC FARMING

Organic farming is a form of agriculture that relies on crop rotation, green manure, compost, biological pest control, and mechanical cultivation to maintain soil productivity and control pests, excluding or strictly limiting the use of synthetic fertilizers and synthetic pesticides, plant growth regulators, livestock feed additives, and genetically modified organisms. Since 1990 the market for organic products has grown at a rapid pace, averaging 20-25 percent per year to reach $33 billion in 2005. This demand has driven a similar increase in organically managed farmland. Approximately 306,000 square kilometres (30.6 million hectares) worldwide are now farmed organically, representing approximately 2% of total world farmland. In addition, as of 2005 organic wild products are farmed on approximately 62 million hectares.

Organic agricultural methods are internationally regulated and legally enforced by many nations, based in large part on the standards set by the International Federation of Organic Agriculture Movements (IFOAM), an international umbrella organization for organic organizations established in 1972. IFOAM defines the overarching goal of organic farming as follows:

"Organic agriculture is a production system that sustains the health of soils, ecosystems and people. It relies on ecological processes, biodiversity and cycles adapted to local conditions, rather than the use of inputs with adverse effects. Organic agriculture combines tradition, innovation and science to benefit the shared environment and promote fair relationships and a good quality of life for all involved.." —International Federation of Organic Agriculture Movements The organic movement began in the early 1930s and early 1940s as a reaction to agriculture's growing reliance on synthetic fertilizers. Artificial fertilizers had been created during the 18th century, initially with superphosphates and then ammonia derived fertilizers mass-produced using the Haber-Bosch process developed during World War I. These early fertilizers were cheap, powerful, and easy to transport in bulk. Similar advances occurred in chemical pesticides in the 1940s, leading to the decade being referred to as the 'pesticide era'.

Sir Albert Howard is widely considered to be the father of organic farming. Further work was done by J.I. Rodale in the United States, Lady Eve Balfour in the United Kingdom, and many others across the world.

As a percentage of total agricultural output, organic farming has remained tiny since its beginning. As environmental awareness and concern increased amongst the general population, the originally supply-driven movement became demand-driven. Premium prices from consumers and in some cases government subsidies attracted many farmers into converting. In the developing world, many farmers farm according to traditional methods which are comparable to organic farming but are not certified. In other cases, farmers in the developing world have converted out of necessity. As a proportion of total global agricultural output, organic output remains small, but it has been growing rapidly in many countries, notably in Europe.

"An organic farm, properly speaking, is not one that uses certain methods and substances and avoids others; it is a farm whose structure is formed in imitation of the structure of a natural system that has the integrity, the independence and the benign dependence of an organism" —Wendell Berry, "The Gift of Good Land"

The term holistic is often used to describe organic farming. Enhancing soil health is the cornerstone of organic farming. A variety of methods are employed, including crop rotation, green manure, cover cropping, application of compost, and mulching. Organic farmers also use certain processed fertilizers such as seed meal, and various mineral powders such as rock phosphate and greensand, a naturally occurring form of potash. These methods help to control erosion, promote biodiversity, and enhance the health of the soil.

Pest control targets animal pests (including insects), fungi,weeds and disease. Organic pest control involves the cumulative effect of many techniques, including, allowing for an acceptable level of pest damage, encouraging or even introducing beneficial organisms, careful crop selection and crop rotation, and mechanical controls such as row covers and traps. These techniques generally provide benefits in addition to pest control—soil protection and improvement, fertilization, pollination, water conservation, season extension, etc.—and these benefits are both complementary and cumulative in overall effect on farm health. Effective organic pest control requires a thorough understanding of pest life cycles and interactions. Most organic farms use less pesticides than most conventional farms. The main three used are Bt (a bacterial toxin), pyrethrum and rotenone. Surveys have found that fewer than 10% of organic farmers use these pesticides regularly; one survey found that only 5.3% of vegetable growers in California use rotenone while 1.7% use pyrethrum (Lotter 2003:26).

Weeds are controlled mechanically, thermically and through the use of covercrops and mulches. The traditional method is to remove weeds by hand, which is still practiced in developing countries by small scale farmers. However, this has proven too costly in developed countries where labor is more expensive. One recent innovation in rice farming is to introduce ducks and fish to wet paddy fields, which eat both weeds and insects.

STANDARDS REGULATE PRODUCTION METHODS

Standards regulate production methods and in some cases final output for organic agriculture. Standards may be voluntary or legislated. As early as the 1970s organic producers could be voluntarily certified by private associations. In the 1980s, governments began to produce organic production guidelines. Beginning in the 1990s, a trend towards legislation of standards began, most notably with the 1991 EU-Eco-regulation developed for European Union, which set standards for 12 countries, and a 1993 UK programme. The EU's programme was followed by a Japan programme in 2001, and in 2002 the United States created the National Organic Programme (NOP). As of 2007 over 60 countries have regulations on organic farming (IFOAM 2007:11). In 2005 IFOAM created the Principles of Organic Agriculture, an international guideline for certification criteria. Typically the agencies do not certify individual farms, but rather accredit certification groups.

Materials used in organic production and foods are tested independently by the Organic Materials Review Institute.

Under USDA organic standards, manure must be composted and allowed to reach a sterilizing temperature. If raw animal manure is used, 120 days must pass before the crop is harvested. The economics of organic farming, a subfield of agricultural economics, encompasses the entire process and effects of organic farming in terms of human society, including social costs, opportunity costs, unintended consequences, information asymmetries, and economies of scale. Although the scope of economics is broad, agricultural economics tends to focus on maximizing yields and efficiency at the farm level.

Mainstream economics takes an anthropocentric approach to the value of the natural world: biodiversity, for example, is considered beneficial only to the extent that it is valued by people and increases profits. Some governments such as the European Union subsidize organic farming, in large part because these countries believe in the external benefits of reduced water use, reduced water contamination by pesticides and nutrients of organic farming, reduced soil erosion, reduced carbon emissions, increased biodiversity, and assorted other benefits. Organic farming is labor and knowledge-intensive whereas conventional farming is capital-intensive, requiring more energy and manufactured inputs. Organic farmers in California have cited marketing as their greatest obstacle.

MARKETS FOR ORGANIC PRODUCTS

The markets for organic products are strongest in North America and Europe, which as of 2001 are estimated to have \$6 and \$8 billion respectively of the \$20 billion market (2003:6). However, as of 2007 organic farmland is distributed across the globe. Australasia has 39% of the total organic farmland with Australia's 11.8 million hectares, but 97 percent of this land is sprawling rangeland (2007:35), which results in total sales of approximately 5% of US sales (2003:7). Europe has 23 percent of total organic farmland (6.9 million hectares), followed by Latin America with 19 percent (5.8 million hectares). Asia has 9.5 percent while North America has 7.2 percent. Africa has a mere 3 percent.

Besides Australia, the countries with the most organic area are Argentina (3.1 million hectares), China (2.3 million hectares), and the United States (1.6 million hectares). Much of Argentina's organic farmland is pasture, like that of Australia (2007:42). Italy, Spain, Germany, Brazil, Uruguay, and the UK follow the United States by the amount of land managed organically.

The estimated total market value of certified organic products was estimated to be \$20 billion. By 2002 this was \$23 billion and by 2005 \$33 billion, with Organic Monitor projecting sales of \$40 billion in 2006. The change from 2001 to 2005 represents a compound growth of 10.6 percent. In recent years both Europe and North America have experienced strong growth in organic farmland. Each added half a million hectares from 2004 to 2007 — for the US this is a 29 percent change. However, this growth has occurred under different conditions. While the European Union has shifted agricultural subsidies to organic farmers in recognition of its environmental benefits, the United States has taken a free market approach. As a result, as of 2001 3 percent of European farmland was organically managed compared to just 0.3 percent of United States farmland. By 2005 Europe's organic land was 3.9 percent while the United States' had risen to 0.6 percent.

IFOAM's The World of Organic Agriculture: Statistics and Emerging Trends 2007 lists the countries which added the most hectares and had the highest percentage growth in 2007. Among these, China is listed third in adding the most hectares behind the United States and Argentina. China jumped from approximately 300,000 hectares of organic land in 2005 to approximately 3.5 million hectares in 2006 — an increase of over a thousand percent. This rise can be attributed to the certification of China's Organic Food Development

Centre in 2002 by IFOAM. The end of 2005 marks the end of the three-year transition period begun in 2002..

A 2006 study suggests that converted organic farms have lower pre-harvest yields than their conventional counterparts in developed countries (92%) and that organic farms have higher pre-harvest yields than their low-intensity counterparts in developing countries (132%). The researcher attributes this to a relative lack of expensive fertilizers and pesticides in the developing world compared to the intensive, subsidy-driven farming of the developed world. Nonetheless, the researcher purposely avoids making the claim that organic methods routinely outperform green-revolution (conventional) methods. This study incorporated a 1990 review of 205 crop comparisons which found that organic crops had 91% of conventional yields. A major US survey published in 2001, analyzed results from 150 growing seasons for various crops and concluded that organic yields were 95-100% of conventional yields.

Lotter reports that repeated studies have found that organic farms withstand severe weather conditions better than conventional farms, sometimes yielding 70-90% more than conventional farms during droughts. A 22-year farm trial study by Cornell University published in 2005 concluded that organic farming produces the same corn and soybean yields as conventional methods over the long-term averages, but consumed less energy and used zero pesticides. The results were attributed to lower yields in general but higher yields during drought years. A study of 1,804 organic farms in Central American hit by Hurricane Mitch in 1998 found that the organic farms sustained the damage much better, retaining 20 to 40% more topsoil and smaller economic losses at highly significant levels than their neighbors.

On the other hand, a prominent 21-year Swiss study found an average of 20% lower organic yields over conventional, along with 50% lower expenditure on fertilizer and energy, and 97% less pesticides. A long-term study by U.S Department of Agriculture Agricultural Research Service (ARS) scientists concluded that, contrary to widespread belief, organic farming can build up soil organic matter better than conventional no-till farming, which suggests long-term yield benefits from organic farming. An 18-year study of organic methods on nutrient-depleted soil concluded that conventional methods were superior for soil fertility and yield in a cold-temperate climate, arguing that much of the benefits from organic farming are derived from imported materials which could not be regarded as "self-sustaining".

While organic farms have lower yields, organic methods require no synthetic fertilizer and pesticides. The decreased cost on those inputs, along with the premiums which consumers pay for organic produce, create higher profits for organic farmers. Organic farms have been consistently found to be as or more profitable than conventional farms with premiums included, but without premiums profitability is mixed. Welsh reports that organic farmers are more profitable in the drier states of the United States, likely due to their superior drought performance.

In 2008 the UN Environmental Programme (UNEP) and UN Conference on Trade and Development (UNCTAD) issued a report which stated that "organic agriculture can be more conducive to food security in Africa than most conventional production systems, and that it is more likely to be sustainable in the long-term". The report assessed 114 projects in 24 African countries, finding that "yields had more than doubled where organic, or near-organic practices had been used" and that soil fertility and drought resistance improved.

ORGANIC METHODS IN MACROECONOMIC IMPACT

Organic methods often require more labor, providing rural jobs but increasing costs to urban consumers.

Agriculture in general imposes external costs upon society through pesticides, nutrient runoff, excessive water usage, and assorted other problems. As organic methods minimize some of these factors, organic farming is believed to impose fewer external costs upon society. A 2000 assessment of agriculture in the UK determined total external costs costs for 1996 of 2343 million British pounds or 208 pounds per hectare. A 2005 analysis of these costs in the USA concluded that cropland imposes approximately 5 to 16 billion dollars ($30 to $96 per hectare), while livestock production imposes 714 million dollars. Both studies concluded that more should be done to internalize external costs, and neither included subsidies in their analysis, but noted that subsidies also influence the cost of agriculture to society. Both focused on purely fiscal impacts. The 2000 review included reported pesticide poisonings but did not include speculative chronic effects of pesticides, and the 2004 review relied on a 1992 estimate of the total impact of pesticides.

Some pesticides may damage human health and the environment, and most organic farms use less pesticides than conventional farms. The main three pesticides used in organic farming are Bt (a bacterial toxin), pyrethrum, rotenone, copper and sulphur. Surveys have found that fewer than 10% of organic farmers use these pesticides regularly; one survey found that only 5.3% of vegetable growers in California use rotenone while 1.7% use pyrethrum. Reduction and elimination of chemical pesticide use is technologically challenging. Few organic farms manage to eliminate the use of pesticides entirely; organic pesticides are often used to compliment other pest control strategies.

Pesticide runoff is one of the most significant effects of pesticide use. The USDA Natural Resources Conservation Service tracks the environmental risk posed by pesticide water contamination from farms, and its conclusion has been that "the Nation's pesticide policies during the last twenty six years have succeeded in reducing overall environmental risk, in spite of slight increases in area planted and weight of pesticides applied. Nevertheless, there are still areas of the country where there is no evidence of progress, and areas where risk levels for protection of drinking water, fish, algae and crustaceans remain high".

Pest resistant genetically modified crops have been proposed as an alternative to pesticide use, however concerns over the safety and the long term benefits of genetically modified food, result in the genetic modification being widely opposed in the organic farming movement.

Food Quality

Organic food is widely believed by the lay public to be healthier, although the research is controversial. Two studies have found that children fed organic diets experienced significantly lower organophosphorus pesticide exposure than children fed conventional diets. Although the researchers did not collect health *outcome* data in this study, they concluded "it is intuitive to assume that children whose diets consist of organic food items would have a lower probability of neurologic health risks". A 2007 study found that consumption of organic milk is associated with a decrease in risk for eczema, although no comparable benefit was found for organic fruits, vegetables, or meat.

Extensive scientific research is being carried out in Switzerland at over 200 farms to determine differences in the quality of organic food products compared to conventional in addition to other tests. The FiBL Institute has been investigating the differences at over 200 farms. It states that "organic products stand out as having higher levels of secondary plant compounds and vitamin C. In the case of milk and meat, the fatty acid profile is often better from a nutritional point of view. As far as carbohydrates and minerals, organic products are no different from conventional products. However, in regard to undesirables such as nitrate and pesticide residues, organic products have a clear advantage. A £12m EU-funded investigation into the difference between organic and ordinary farming published in 2007 found that organic foods have more nutritional value. A recent study found that organically grown produce has double the flavonoids, an important antioxidant.. A 2007 study found that organically grown kiwifruit had more antioxidants than conventional kiwifruit.

Genetically Modified Organisms

A key characteristic of organic farming is rejection of genetically engineered products, including plants and animals. On October 19, 1998, participants at IFOAM's 12th Scientific Conference of IFOAM) issued the Mar del Plata Declaration, where more than 600 delegates from over 60 countries voted unanimously to exclude the use of genetically modified organisms in food production and agriculture. From this point, it became widely recognized that GMOs are categorically excluded from organic farming. Although opposition to the use of any transgenic technologies in organic farming is strong, agricultural researchers Luis Herrera-Estrella & Ariel Alvarez-Morales continue to advocate integration of transgenic technologies into organic farming as the optimal means to sustainable agriculture, particularly in the developing world. Similarly, some organic farmers question the rationale behind the ban on the use of genetically engineered seed because they see it a biological technology consistent with organic principles

Although GMOs are excluded from use in organic farming, there is concern that the pollen from genetically modified crops is increasingly contaminating organic and heirloom genetics making it difficult, if not impossible, to keep these genetics from entering the organic food supply. International trade restrictions limit the availability GMOs to certain countries. The actual dangers that genetic modification could pose to the environment or, supposedly, individual health, are hotly contended.

Soil Conservation

In *Dirt: The Erosion of Civilizations*, geomorphologist David Montgomery outlines a coming crisis from soil erosion. Agriculture relies on roughly one meter of topsoil, and that is being depleted ten times faster than it is being replaced. No-till farming, which some claim depends upon pesticides, is regarded as one way to minimize erosion. However, a recent study by the USDA's Agricultural Research Service has found that organic farming is even better at building up the soil than no-till.

The elimination of synthetic nitrogen in organic systems decreases fossil fuel consumption by 33 percent (LaSalle)and carbon sequestration takes CO_2 out of the atmosphere by putting it in the soil in the form of organic matter which is often lost in conventionally managed soils. Carbon sequestration occurs at especially high levels in organic no-till managed soil according to the Rodale Institute.

Nutrient Leaching

Excess nutrients in lakes, rivers, and groundwater can cause algal blooms, eutrophication, and subsequent dead zones. In addition, nitrates are harmful to aquatic organisms by themselves. The main contributor to this pollution is nitrate fertilizers whose use is expected to "double or almost triple by 2050". Researchers at the United States National Academy of Sciences found that that organically fertilizing fields "significantly [reduces] harmful nitrate leaching" over conventionally fertilized fields: "annual nitrate leaching was 4.4-5.6 times higher in conventional plots than organic plots".

Scientists believe that the large dead zone in the Gulf of Mexico is caused in large part by agricultural pollution: a combination of fertilizer runoff and livestock manure runoff. A study by the United States Geological Survey (USGS)

found that over half of the nitrogen released into the Gulf comes from agriculture. The economic cost of this for fishermen may be large, as they must travel far from the coast to find fish.

At the 2000 IFOAM Conference, researchers presented a study of nitrogen leaching into the Danube River. They found that nitrogen runoff was substantially lower among organic farms and suggested that the external cost could be internalized by charging 1 euro per kg of nitrogen released. A 2005 study found a strong link between agricultural runoff and algae blooms in California.

Sales and Marketing

Organic farmers report that marketing and distribution are difficult obstacles. Most of organic sales are concentrated in developed nations. These products are what economists call credence goods in that they rely on uncertain certification. As food prices rise, organic products may experience falling demand. A 2008 survey by WSL Strategic Retail found that interest in organic products had dropped since 2006, and that 42% of Americans polled don't trust organic produce. and The Hartman Group reports that 69% of Americans claim to occasionally buy organic products, down from 73% in 2005. The Hartman Group says that people may be substituting local produce for organic produce.

In the United States, 75% of organic farms are smaller than 2.5 hectares and in California 2% of the farms account for over half of the sales. Groups of small farms join together in cooperatives such as Organic Valley, Inc. to market their goods more effectively.

Over the past twenty years, however, most of these cooperative distributors have merged or been bought out. Rural sociologist Philip H. Howard has researched the structure and transformation of the organic industry in the United States. He claims that in 1982 there were 28 consumer cooperative distributors but as of 2007 there are only 3, and he has created a graphic displaying the consolidation. His research shows that most of these small cooperatives have been absorbed into large multinational corporations such as General Mills, Heinz, ConAgra, Kellog, and assorted other brands. This consolidation has raised concerns among consumers and journalists of potential fraud and degradation in standards. Most of these large corporations sell their organic products through subsidiaries, allowing them to keep their names off the labels.

Biodiversity

A wide range of organisms benefit from organic farming, but it is unclear whether organic methods confer greater benefits than integrated agri-environmental conventional programmes. Nearly all non-crop, naturally-occurring species observed in comparative farm land practice studies show a preference in organic farming both by population and richness. Spanning all associated species, there is an average of 30% more on organic farms versus conventional farming methods. Birds, butterflies, soil microbes, beetles, earthworms, spiders, vegetation, and mammals are particularly affected.

Organic crops use little or no herbicides and pesticides and thus biodiversity fitness and population density benefit. Many weed species attract beneficial insects that improve soil qualities and forage on weed pests. Soil-bound organisms often benefit because of increased bacteria populations due to natural fertilizer spread such as manure, while experiencing reduced intake of herbicides and pesticides commonly associated with conventional farming methods.

Increased biodiversity, especially from soil microbes such as mycorhizzae, have been proposed as an explanation for the high yields experienced by some organic plots, especially in light of the differences seen in a 21-year comparison of organic and control fields.

The level of biodiversity that can be yielded from organic farming provides a natural capital to humans. Species found in most organic farms provides a means of agricultural sustainability by reducing amount of human input (*e.g.* fertilizers, pesticides). Farmers that produce with organic methods reduce risk of poor yields by promoting biodiversity.

Common game birds such as the ring-necked pheasant and the northern bobwhite often reside in agriculture landscapes, and are a natural capital yielded from high demands of recreational hunting. Because bird species richness and population are typically higher on organic farm systems, promoting biodiversity can be seen as logical and economical.

Biological research on soil and soil organisms has proven beneficial to the system of organic farming. Varieties of bacteria and fungi break down chemicals, plant matter and animal waste into productive soil nutrients. In turn, the producer benefits by healthier yields and more arable soil for future crops. Furthermore, a 21-year study was conducted testing the effects of organic soil matter and its relationship to soil quality and yield.

Controls included actively managed soil with varying levels of manure, compared to a plot with no manure input. After the study commenced, there was significantly lower yields on the control plot when compared to the fields with manure. The concluded reason was an increased soil microbe community in the manure fields, providing a healthier, more arable soil system.

Farmers' Markets

Price premiums are important for the profitability of small organic farmers, and so many sell directly to consumers in farmers' markets. In the United States the number of farmers' markets has grown from 1,755 in 1994 to 4,385 in 2006.

Capacity Building

Organic agriculture can contribute to meaningful socio-economic and ecologically sustainable development, especially in poorer countries. On one hand, this is due to the application of organic principles, which means efficient management of local resources (*e.g.* local seed varieties, manure, etc.) and therefore cost-effectiveness. On the other hand, the market for organic products – at local and international level – has tremendous growth prospects and offers creative producers and exporters in the South excellent opportunities to improve their income and living conditions.

Organic Agriculture is a very knowledge intensive production system. Therefore capacity building efforts play a central role in this regard. There are many efforts all around the world regarding the development of training material and the organization of training courses related to Organic Agriculture. Big parts of existing knowledge is still scattered and not easy accessible. Especially in Developing Countries this situation remains an important constraint for the growth of the organic sector.

For that reason, the International Federation of Organic Agriculture Movements created an Internet Training Platform whose objective is to become the global reference point for Organic Agriculture training through free access to high quality training materials and training programmes on Organic Agriculture. In November 2007, the Training Platform hosted more than 170 free manuals and 75 training opportunities.

Organic Aagricultural Systems

A number of critics contest the notion that organic agricultural systems are more friendly to the environment and more sustainable than high-yielding farming systems. Among these critics are Norman Borlaug, father of the "green revolution," and winner of the Nobel Peace Prize, who asserts that organic farming practices can at most feed 4 billion people, after expanding cropland dramatically and destroying ecosystems in the process and Prof A. Trewavas.

The debate has been summarized in an exchange between Trewavas and Lord P. Melchett, and published by a major supermarket, concerned about examining the issues.

One study from the Danish Environmental Protection Agency found that, area-for-area, organic farms of potatoes, sugar beet and seed grass produce as little as half the output of conventional farming.

In 2008 a study from UN Environmental Programme concluded that organic methods greatly increase yields in Africa and a review of over two hundred crop comparisons argued that organic farming could produce enough food per capita to sustain the current human population; the difference in yields between organic and non-organic methods were small, with non-organic methods resulting in slightly higher yields in developed areas and organic methods resulting in slightly higher yields in developing areas. That analysis has been severely criticised by Alex Avery, who contends that the review claimed many non-organic studies to be organic, misreported organic yields, made false comparisons between yields of organic and non-organic studies which were not comparable, counted high organic yields several times by citing different papers which referenced the same data, and gave equal weight to studies from sources which were not impartial and rigorous university studies.

Urs Niggli, director of the FiBL Institute contents that the wave of newspaper articles like 'Organic food exposed' or 'The hypocrisy of organic farmers' are a part of a global campaign against organic farming that take their arguments mostly from the book 'The truth about organic farming', by Alex Avery of the Hudson Institute.

In 1998, Dennis Avery of the Hudson Institute claimed the risk of E. coli infection was eight times higher when eating organic food rather than non-organic food, using the Centre for Disease Control (CDC) as a source. When the CDC was contacted, it stated that there was no evidence for the claim. The *New York Times* commented on Avery's attacks: "The attack on organic food by a well-financed research organization suggests that, though organic food accounts for only 1 percent of food sales in the United States, the conventional food industry is worried."

Organic agriculture can reduce the level of negative externalities from (conventional) agriculture. Whether this is seen as private or public benefits depends upon the initial specification of property rights. However, it is clear that agriculture has been undervalued and underestimated as a means to combat global climate change. Soil carbon data recorded by The Rodale Institute show that regenerative organic agricultural practices are among the most effective strategies for mitigating CO_2 emissions.

3

Indian Agriculture in a Historical Context

An important aspect to note about Indian agricultural production (in most parts of the country) is its relationship (even dependency) on skilful and wise water-management practices. One of the unique features of the Indian Subcontinent is how almost its entire rainfall is concentrated in the few monsoon months. During the monsoons, the Indian Subcontinent is usually gifted with bountiful rains, although not infrequently, this bountiful monsoon can turn into a terror, causing uncontrollable floods in parts of the country. Conversely, every few years, the monsoon can be erratic and deficient, leading to drought and the possibility of famine.

Thus. all through history the development of Indian agriculture has been inextricably linked with effective water-management practices that have either been taken up by the state, or by local village communities. Water management has necessitated a certain degree of cooperation and collective spirit in the Indian countryside, and until the imposition of colonial rule, it precluded any widespread development of private property in India.

Regional rulers, or local representatives of the state were generally obliged to allocate a certain percentage of the agricultural taxes on building and managing water-storage, water-harvesting and/or water-diverting structures which facilitated a second crop, and provided water for drinking and other purposes in the long dry season.

Only a small percentage of Indian farmers have enjoyed the luxury of natural irrigation, although there are reports that in certain parts of the country, the soil used to retain enough moisture well beyond the monsoon months. However, it is equally true that the drying up of wells led to mass migrations, and sudden depopulation of old towns and villages.

In Assam, Bengal and Bihar - all flood-prone states, there is evidence of a massive network of canals that allowed both effective drainage to prevent flooding during the heavy monsoon months, and also provide for fishing, transportation and irrigation arteries in the dry seasons.

Intelligent water-management thus allowed for the growth of a healthy agricultural surplus, that in turn facilitated steady urbanization (albeit at a much slower pace than seen in the industrial era), and the development of a variety of pre-industrial manufacturing in areas such as textiles, jewelry, wood/metal-working etc. Because of the immense importance of effective water-management, the entire revenue system of the state was structured so as to take into account both the necessity of water-management and the inherent dependency that existed between Indian agriculture and the availability of water. Most Indian states attempted to collect revenues in a manner that did not entirely destroy the village solidarity that was essential in proper sharing and management of common water-resources. At the same time, urban tax collectors had to deal with mediating entities from the villages, so that they did not tax at a rate that might lead to the destruction of water-management facilities so essential for life and sustainable agriculture.

By and large, taxes were imposed on villages collectively (not on individual farmers directly), and the village elites (whether Brahmins or others) were obliged to ensure that the burden of taxes did not destroy the complete viability of agriculture. Taxes were also adjusted keeping in mind whether the land was well-irrigated or not. When ruling dynasties became utterly corrupt and decadent, and failed to tax within limits, or neglected the construction and maintenance of collective water-management facilities, they were confronted with revolts and rebellions, and even risked being overthrown. There is evidence that whenever the tax collectors of a certain regime became too greedy (such as what happened in certain villages/districts during the reigns of some of Northern India's more rapacious invading conquerors), entire villages would be abandoned to escape the burden of heavy taxes. As long as India was not very densely populated and there was the potential of new lands to cultivate, (and the reach of the armed state was not absolute and all-encompassing), this was one of the means by which Indian villagers - both artisans and peasants were able to escape the burden of back-breaking taxes.

But as the population grew and states became more powerful, this escape route was gradually sealed off, and this then led to regional peasant rebellions - most significant of which were the Sikh and the Maratha revolts against the Mughals. Thus, the British arrived in India at a time when peasant rebellions were beginning to eat away at the invincibility of the might of the Mughal empire. However, even as the decentralization that resulted from the break-up of the widely resented Mughal empire eased slightly the burden on the peasantry in most instances, it also provided an opportunity for the increasingly cunning and shrewd Britishers in India to seize the opportunity, and exploit the widening cleavages in Indian society for their own nefarious ends.

HISTORY OF AGRICULTURE IN THE INDIA

Indian agriculture began by 9000 BCE as a result of early cultivation of plants, and domestication of crops and animals. Settled life soon followed with implements and techniques being developed for agriculture. Double monsoons led to two harvests being reaped in one year. Indian products soon reached the world via existing trading networks and foreign crops were introduced to India. Plants and animals—considered essential to their survival by the Indians—came to be worshiped and venerated.

The middle ages saw irrigation channels reach a new level of sophistication in India and Indian crops affecting the economies of other regions of the world under Islamic patronage. Land and water management systems were developed with an aim of providing uniform growth. Despite some stagnation during the later modern era the independent Republic of India was able to develop a comprehensive agricultural programme.

EARLY HISTORY

Wheat, barley and jujube were domesticated in the Indian subcontinent by 9000 BP. Domestication of sheep and goat soon followed. This period also saw the first domestication of the elephant. Barley and wheat cultivation—along with the domestication of cattle, primarily sheep and goat—was visible in Mehrgarh by 8000-6000 BCE. Agro pastoralism in India included threshing, planting crops in rows—either of two or of six—and storing grain in granaries. In the period of the Neo-lithic revolution (roughly 8000-5000 BCE.), agriculture was far from the dominant mode of support for human societies. But those who adopted it, have survived and increased, and passed their techniques of production to the next gene4ration.

This transformation of knowledge was the base of further development in agriculture. By the 5th millennium BCE agricultural communities became widespread in Kashmir. Zaheer Baber (1996) writes that 'the first evidence of cultivation of cotton had already developed'. Cotton was cultivated by the 5th millennium BCE-4th millennium BCE. The Indus cotton industry was well developed and some methods used in cotton spinning and fabrication continued to be practiced till the modern Industrialisation of India.

A variety of tropical fruit such as mango and muskmelon are native to the Indian subcontinent. The Indians also domesticated hemp, which they used for a number of applications including making narcotics, fibre, and oil. The farmers of the Indus Valley, which thrived in modern-day Pakistan and North India, grew peas, sesame, and dates. Sugarcane was originally from tropical South Asia and Southeast Asia. Different species likely originated in different locations with *S. barberi* originating in India and *S. edule* and *S. officinarum* coming from New Guinea. Wild Oryza rice appeared in the Belan and Ganges valley regions of northern India as early as 4530 BCE and 5440 BCE respectively. Rice was cultivated in the Indus Valley Civilization. Agricultural activity during the second millennium BC included rice cultivation in the Kashmir and Harrappan regions. Mixed farming was the basis of the Indus valley economy.

Denis J. Murphy details the spread of cultivated rice from India into South-east Asia:

- Several wild cereals, including rice, grew in the Vindhyan Hills, and rice cultivation, at sites such as Chopani-Mando and Mahagara, may have been underway as early as 7000 BP. The relative isolation of this area and the early development of rice farming imply that it was developed indigenously....Chopani-Mando and Mahagara are located on the upper reaches of the Ganges drainage system and it is likely that migrants from this area spread rice farming down the Ganges valley into the fertile plains of Bengal, and beyond into south-east Asia.

Irrigation was developed in the Indus Valley Civilization by around 4500 BCE. The size and prosperity of the Indus civilization grew as a result of this innovation, which eventually led to more planned settlements making use of drainage and sewers. Sophisticated irrigation and water storage systems were developed by the Indus Valley Civilization, including artificial reservoirs at Girnar dated to 3000 BCE, and an early canal irrigation system from circa 2600 BCE. Archeological evidence of an animal-drawn plough dates back to 2500 BC in the Indus Valley Civilization.

VEDIC PERIOD – POST MAHA JANAPADAS PERIOD (1500 BCE – 200 CE)

- Gupta finds it likely that summer monsoons may have been longer and may have contained moisture in excess than required for normal food production. One effect of this excessive moisture would have been to aid the winter monsoon rainfall required for winter crops. In India, both wheat and barley are held to be *Rabi* (winter) crops and—like other parts of the world—would have largely depended on winter monsoons before the irrigation became widespread. The growth of the *Kharif* crops would have probably suffered as a result of excessive moisture. Jute was first cultivated in India, where it was used to make ropes and cordage. Some animals—thought by the Indians as being vital to their survival—came to be worshiped. Trees were also domesticated, worshiped, and venerated—*Pipal* and *Banyan* in particular. Others came to be known for their medicinal uses and found mention in the holistic medical system *Ayurveda*. The Encyclopædia Britannica—on the subject of agriculture of the later Vedic period—holds that:In the later Vedic texts (c. 1000–500 BC), there are repeated references to iron. Cultivation of a wide range of cereals, vegetables, and fruits is described. Meat and milk products were part of the diet; animal husbandry was important. The soil was plowed several times. Seeds were broadcast. Fallowing and a certain sequence of cropping were recommended. Cow dung provided the manure. Irrigation was practiced.

- The Mauryan Empire (322–185 BCE) categorised soils and made meteorological observations for agricultural use. Other Mauryan facilitation included construction and maintenance of dams, and provision of horse-drawn chariots—quicker than traditional bullock carts. The Greek diplomat Megasthenes (c. 300 BC)—in his book *Indika*— provides a secular eyewitness account of Indian agriculture:India has many huge mountains which abound in fruit-trees of every kind, and many vast plains of great fertility.... The greater part of the soil, moreover, is under irrigation, and consequently bears two crops in the course of the year.... In addition to cereals, there grows throughout India much millet... and much pulse of different sorts, and rice also, and what is called bosporum Indian millet.... Since there is a double rainfall *i.e.*, the two monsoons in the course of each year... the inhabitants of India almost always gather in two harvests annually.

EARLY COMMON ERA – HIGH MIDDLE AGES (200–1200 CE)

The Tamil people cultivated a wide range of crops such as rice, sugarcane, millets, black pepper, various grains, coconuts, beans, cotton, plantain, tamarind and sandalwood. Jackfruit, coconut, palm, areca and plantain trees were also known. Systematic ploughing, manuring, weeding, irrigation and crop protection was practiced for sustained agriculture. Water storage systems were designed during this period. Kallanai (1st-2nd century CE), a dam built on river Kaveri during this period, is considered the as one of the oldest water-regulation structures in the world still in use.

Spice trade involving spices native to India—including cinnamon and black pepper—gained momentum as India starts shipping spices to the Mediterranean. Roman trade with India followed as detailed by the archaeological record and the *Periplus of the Erythraean Sea*. Chinese sericulture attracted Indian sailors during the early centuries of the common era. Crystallised sugar was discovered by the time of the Guptas (320-550 CE), and the earliest reference of candied sugar come from India. The process was soon transmitted to China with traveling Buddhist monks. Chinese documents confirm at least two missions to India, initiated in 647 CE, for obtaining technology for sugar-refining. Each mission returned with results on refining sugar.

Indian spice exports find mention in the works of:

- Ibn Khurdadhbeh (850),
- Al-Ghafiqi (1150),

- Ishak bin Imaran (907) and
- Al Kalkashandi (fourteenth century).

Noboru Karashima's research of the agrarian society in South India during the Chola Empire (875-1279) reveals that during the Chola rule land was transferred and collective holding of land by a group of people slowly gave way to individual plots of land, each with their own irrigation system. The growth of individual disposition of farming property may have led to a decrease in areas of dry cultivation. The Cholas also had bureaucrats which oversaw the distribution of water—particularly the distribution of water by tank-and-channel networks to the drier areas.

LATE MIDDLE AGES – EARLY MODERN ERA (1200–1757 CE)

- The construction of water works and aspects of water technology in India is described in Arabic and Persian works. The diffusion of Indian and Persian irrigation technologies gave rise to an irrigation systems which brought about economic growth and growth of material culture. Agricultural 'zones' were broadly divided into those producing rice, wheat or millets. Rice production continued to dominate Gujarat and wheat dominated north and central India. The Encyclopædia Britannica details the many crops introduced to India during this period of extensive global discourse:Introduced by the Portuguese, cultivation of tobacco spread rapidly. The Malabâr Coast was the home of spices, especially black pepper, that had stimulated the first European adventures in the East. Coffee had been imported from Abyssinia and became a popular beverage in aristocratic circles by the end of the century. Tea, which was to become the common man's drink and a major export, was yet undiscovered, though it was growing wild in the hills of Assam. Vegetables were cultivated mainly in the vicinity of towns. New species of fruit, such as the pineapple, papaya, and cashew nut, also were introduced by the Portuguese. The quality of mango and citrus fruits was greatly improved.

Land management was particularly strong during the regime of Akbar the Great (reign: 1556-1605), under whom scholar-bureaucrat Todarmal formulated and implemented elaborated methods for agricultural management on a rational basis.

Indian crops—such as cotton, sugar, and citric fruits—spread visibly throughout North Africa, Islamic Spain, and the Middle East.

Though they may have been in cultivation prior to the solidification of Islam in India, their production was further improved as a result of this recent wave, which led to far-reaching economic outcomes for the regions involved.

COLONIAL BRITISH ERA (1757–1947 CE)

- Few Indian commercial crops—such as Cotton, indigo, opium, and rice—made it to the global market under the British Raj in India. The second half of the 19th century saw some increase in land under cultivation and agricultural production expanded at an average rate of about 1 percent per year by the later 19th century. Due to extensive irrigation by canal networks Punjab, Narmada valley, and Andhra Pradesh became centers of agrarian reforms. Roy (2006) comments on the Influence of the world wars on the Indian agricultural system:Agricultural performance in the interwar period (1918–1939) was dismal. From 1891 to 1946, the annual growth rate of all crop output was 0.4 percent, and food-grain output was practically stagnant. There were significant regional and intercrop differences, however, non-food crops doing better than food crops. Among food crops, by far the most important source of stagnation was rice. Bengal had below-average growth rates in both food and non-food crop output, whereas Punjab and Madras were the least stagnant regions. In the interwar period, population growth accelerated while food output decelerated, leading to declining availability of food per head. The crisis was most acute in Bengal, where food output declined at an annual rate of about 0.7 percent from 1921 to 1946, when population grew at an annual rate of about 1 percent.

The British regime in India did supply the irrigation works but rarely on the scale required. Community effort and private investment soared as market for irrigation developed. Agricultural prices of some commodities rose to about three times between 1870-1920.

A rich source of the state of Indian agriculture in the early British era is a report prepared by a British engineer, Thomas Barnard, and his Indian guide, Raja Chengalvaraya Mudaliar, around 1774. This report contains data of agricultural production in about 800 villages in the area around Chennai in the years 1762 to 1766. This report is available in Tamil in the form of palm leaf manuscripts at Thanjavur Tamil University, and in English in the Tamil Nadu State Archives. A series of articles in The Hindu newspaper in the early 1990s authored by researchers at The Center for Policy Studies led by Shri Dharampal highlight the impressive production statistics of Indian farmers of that era.

REPUBLIC OF INDIA (1947 CE ONWARDS)

Special programmes were undertaken to improve food and cash crops supply. The Grow More Food Campaign (1940s) and the Integrated Production Programme (1950s) focused on food and cash crops supply respectively. Five-year plans of India—oriented towards agricultural development—soon followed. Land reclamation, land development, mechanisation, electrification, use of chemicals—fertilizers in particular, and development of agriculture oriented 'package approach' of taking a set of actions instead of promoting single aspect soon followed under government supervision. The many 'production revolutions' initiated from 1960s onwards included Green Revolution in India, Yellow Revolution (oilseed: 1986-1990), Operation Flood (dairy: 1970-1996), and Blue Revolution (fishing: 1973-2002) etc. Following the economic reforms of 1991, significant growth was registered in the agricultural sector, which was by now benefiting from the earlier reforms and the newer innovations of Agro-processing and Biotechnology.

Due to the growth and prosperity that followed India's economic reforms a strong middle class emerged as the main consumer of fruits, dairy, fish, meat and vegetables—a marked shift from the earlier staple based consumption. Since 1991, changing consumption patterns led to a 'revolution' in 'high value' agriculture while the need for cereals is experienced a decline. The per capita consumption of cereals declined from 192 to 152 kilograms from 1977 to 1999 while the consumption of fruits increased by 553 per cent, vegetables by 167 per cent, dairy products by 105 per cent, and non-vegetarian products by 85 per cent in India's rural areas alone. Urban areas experienced a similar increase. Agricultural exports continued to grow at well over 10.1 per cent annually through the 1990s. Contract farming—which requires the farmers to produce crops for a company under contract—and high value agricultural product increased. Contract farming led to a decrease in transaction costs while the contract farmers made more profit compared to the non-contract workforce. However, small landholding continued to create problems for India's farmers as the limited land resulted in limited produce and limited profits.

Since independence, India has become one of the largest producers of wheat, edible oil, potato, spices, rubber, tea, fishing, fruits, and vegetables in the world. The Ministry of Agriculture oversees activities relating to agriculture in India. Various institutions for agriculture related research in India were organised under the Indian Council of Agricultural Research (est. 1929). Other organisations such as the National Dairy Development Board (est. 1965), and National Bank for Agriculture and Rural Development (est. 1982) aided the formation of cooperatives and improved financing.

The contribution of agriculture in employing India's male workforce declined from 75.9 per cent in 1961 to 60 per cent in 1999–2000. Dev holds that 'there were about 45 million agricultural labour households in the country in 1999–2000.' These households recorded the highest incidence of poverty in India from 1993 to 2000. The green revolution introduced high yielding varieties of crops which also increased the usage of fertilizers and pesticides. About 90 per cent of the pesticide usage in India is accounted for by DDT and Lindane (BHC/HCH). There has been a shift to organic agriculture particularly for exported commodities. During 2003-04, agriculture accounted for 22 per cent of India's GDP and employed 58 per cent of the country's workforce. India is the world's largest producer of milk, fruits, cashew nuts, coconuts, ginger, turmeric, banana, sapota, pulses, and black pepper. India is the second largest producer of groundnut, wheat, vegetables, sugar and fish in the world. India is also the third largest producer of tobacco and rice, the fourth largest producer of coarse grains, the fifth largest producer of eggs, and the seventh largest producer of meat.

IRRIGATION IN INDIA

Irrigation in India refers to the supply of water from Indian rivers, tanks, wells, canals and other artificial projects for the purpose of cultivation and agricultural activities. In country such as India, 64 per cent of cultivated land is dependent on monsoons. The economic significance of irrigation in India is namely, to reduce over dependence on monsoons, advanced agricultural productivity, bringing more land under cultivation, reducing instability in output levels, creation of job opportunities, electricity and transport facilities, control of floods and prevention of droughts.

POTENTIAL CAPACITY

India has an irrigation potential of 139.89 million hectares, out of which a minimal 108.2 million hectares(77.35 per cent) of the total land that can be irrigated has been utilised. Currently, about 30 per cent of the net cultivated area has benefited from the irrigation projects that have been implemented. A sum of ₹16,590 crore has been spent of irrigation development up to the 7th Five-Year plans of India. The 10th and 11th Five Year Plan have proposed to invest a sum of ₹1,03,315 crore and 2,10,326 crore on irrigation and flood control in India.

PROJECT CLASSIFICATION

Irrigation Projects in India are classified on two major aspects into:

- Minor Irrigation Projects.
- Medium Irrigation Projects.
- Major Irrigation Projects Since 1950, irrigation works were classified on the basis of cost incurred for the projects' implementation, governing and diss-emination., However, the Planning Commission of India adopted the classification of projects on the basis of culturable command area(CCA).

ACCOMPLISHMENTS

As of 2011, India had a large and diverse agricultural sector, accounting, on average, for about 16 percent of GDP and 10 percent of export earnings. India's arable land area of 159.7 million hectares (394.6 million acres) is the second largest in the world, after the United States. Its gross irrigated crop area of 82.6 million hectares (215.6 million acres) is the largest in the world.

India has grown to become among the top three global producers of a broad range of crops, including wheat, rice, pulses, cotton, peanuts, fruits, and vegetables. Worldwide, as of 2011, India had the largest herds of buffalo and cattle, is the largest producer of milk, and has one of the largest and fastest growing poultry industries.

The following table presents the twenty most important agricultural products in India, by economic value, in 2009. Included in the table is the average productivity of India's farms for each produce. For context and comparison, included is the average of the most productive farms in the world and name of country where the most productive farms existed in 2010. The table suggests India has large potential for further accomplishments from productivity increases, in increased agricultural output and agricultural incomes.

Table. Agriculture in India, Largest Crops by Economic Value.

Value Rank	Produce	Economic (2009 prices, US$)	Unit Price (US$/ kilogram)	Average Yield, India (2010) (tons per Hectare)	World's Most Productive Farms (2010) (tons per Hectare)	Country
1	Rice	$38.42 billion	0.27	3.3	10.8	Australia
2	Buffalo milk	$24.86 billion	0.4	1.7	1.9	Pakistan
3	Cow milk	$17.13 billion	0.31	1.2	10.3	Israel
4	Wheat	$12.14 billion	0.15	2.8	8.9	Nether-lands
5	Mangoes	$9 billion	0.6	6.3	40.6	Cape Verde
6	Sugar cane	$8.92 billion	0.03	66	125	Peru
7	Bananas	$8.38 billion	0.28	37.8	59.3	Indonesia
8	Cotton	$8.13 billion	1.43	1.6	4.6	Israel
9	Fresh Vegetables	$5.97 billion	0.19	13.4	76.8	USA
10	Potatoes	$5.67 billion	0.15	19.9	44.3	USA
11	Tomatoes	$4.59 billion	0.37	19.3	524.9	Belgium
12	Buffalo meat	$4 billion	2.69	0.138	0.424	Thailand
13	Soyabean	$3.33 billion	0.26	1.1	3.7	Turkey
14	Onions	$3.17 billion	0.21	16.6	67.3	Ireland
15	Chicken Meat	$3.12 billion	0.64	10.6	20.2	Cyprus
16	Chick peas	$3.11 billion	0.4	0.9	2.8	China
17	Okra	$3.07 billion	0.35	7.6	23.9	Israel
18	Cattle Meat	$2.93 billion	0.83	13.8	24.7	Jordan
19	Eggs	$2.80 billion	2.7	0.1	0.42	Japan
20	Beans	$2.57 billion	0.42	1.1	5.5	Nicaragua

The Statistics Office of the Food and Agriculture Organisation reported that, per final numbers for 2009, India had grown to become the world's largest producer of the following agricultural produce:

- Fresh Fruit.
- Lemons and limes.
- Buffalo milk, whole, fresh.

- Castor oil seeds.
- Sunflower seeds.
- Sorghum.
- Millet.
- Spices.
- Okra.
- Jute.
- Beeswax.
- Bananas.
- Mangoes, mangosteens, guavas.
- Pulses.
- Indigenous Buffalo Meat.
- Fruit, tropical.
- Ginger.
- Chick peas.
- Areca nuts.
- Other Bastfibres.
- Pigeon peas.
- Papayas.
- Chillies and peppers, dry.
- Anise, badian, fennel, coriander.
- Goat milk, whole, fresh.
- Per final numbers for 2009, India is the world's second largest producer of the following agricultural produce:Wheat.
- Rice.
- Vegetables, fresh.
- Sugar cane.
- Groundnuts, with shell.
- Lentils.
- Garlic.
- Cauliflowers and broccoli.
- Peas, green.
- Sesame seed.
- Cashew nuts, with shell.
- Silk-worm cocoons, reelable.
- Cow milk, whole, fresh.
- Tea.
- Potatoes.
- Onions.
- Cotton lint.
- Cottonseed.
- Eggplants (aubergines).

- Nutmeg, mace and cardamoms.
- Indigenous Goat Meat.
- Cabbages and other brassicas.
- Pumpkins, squash and gourds.

In 2009, India was the world's third largest producer of eggs, oranges, coconuts, tomatoes, peas and beans. In addition to growth in total output, agriculture in India has shown an increase in average agricultural output per hectare in last 60 years.

The table presents average farm productivity in India over three farming years for some crops. Improving road and power generation infrastructure, knowledge gains and reforms has allowed India to increase farm productivity between 40 per cent to 500 per cent over 40 years. India's recent accomplishments in crop yields while being impressive, are still just 30 per cent to 60 per cent.

India and China are competing to establish the world record on rice yields. Yuan Longping of China National Hybrid Rice Research and Development Center, China, set a world record for rice yield in 2010 at 19 tonnes per hectare in a demonstration plot.

Table. Agriculture Productivity in India, Growth in Average Yields from 1970 to 2010.

Crop	Average Yield, 1970-1971 Kilogram per Hectare	Average Yield, 1990-1991 Kilogram per Hectare	Average Yield, 2010–2011 Kilogram per Hectare
Rice	1123	1740	2240
Wheat	1307	2281	2938
Pulses	524	578	689
Oilseeds	579	771	1325
Sugarcane	48322	65395	68596
Tea	1182	1652	1669
Cotton	106	225	510

In 2011, this record was surpassed by an Indian farmer, Sumant Kumar, with 22.4 tonnes per hectare in Bihar, also in a demonstration plot. Both these farmers claim to have employed newly developed rice breeds and System of Rice Intensification (SRI), a recent innovation in rice farming. The claimed Chinese and Indian yields have yet to be demonstrated on 7 hectare farm lots and that these are reproducible over two consecutive years on the same farm.

PROBLEMS

- "Slow agricultural growth is a concern for policymakers as some two-thirds of India's people depend on rural employment for a living. Current agricultural practices are neither economically nor environmentally sustainable and India's yields for many agricultural commodities are low. Poorly maintained irrigation systems and almost universal lack of good extension services are among the factors responsible. Farmers' access to markets is hampered by poor roads, rudimentary market infrastructure, and excessive regulation."—World Bank: "India Country Overview 2008"

- "With a population of just over 1.2 billion, India is the world's largest democracy. In the past decade, the country has witnessed accelerated economic growth, emerged as a global player with the world's fourth largest economy in purchasing power parity terms, and made progress towards achieving most of the Millennium Development Goals. India's integration into the global economy has been accompanied by impressive economic growth that has brought significant economic and social benefits to the country. Nevertheless, disparities in income and human development are on the rise. Preliminary estimates suggest that in 2009-10 the combined all India poverty rate was 32 per cent compared to 37 per cent in 2004-05. Going forward, it will be essential for India to build a productive, competitive, and diversified agricultural sector and facilitate rural, non-farm entrepreneurship and employment. Encouraging policies that promote competition in agricultural marketing will ensure that farmers receive better prices."—World Bank: "India Country Overview 2011

A 2003 analysis of India's agricultural growth from 1970 to 2001, by Food and Agriculture Organisation of the United Nations, identified systemic problems in Indian agriculture. For food staples, the annual growth rate in production during the six-year segments 1970-76, 1976–82, 1982–88, 1988–1994, 1994-2000 were found to be respectively 2.5, 2.5, 3.0, 2.6, and 1.8 percent per annum. Corresponding analyses for the index of total agricultural production show a similar pattern, with the growth rate for 1994-2000 attaining only 1.5 percent per annum.

The low growth rates may constitute in part a response to inadequate returns to Indian farmers. India has very poor rural roads affecting timely supply of inputs and timely transfer of outputs from Indian farms, inadequate irrigation systems, crop failures in some parts of the country because of lack of water while in other parts because of regional floods, poor seed quality and inefficient farming practices in certain parts of India, lack of cold storage and harvest spoilage causing over 30 per cent of farmer's produce going to waste, lack of organised retail and competing buyers thereby limiting Indian farmer's ability to sell the surplus and commercial crops.

The Indian farmer receives just 10 to 23 percent of the price the Indian consumer pays for exactly the same produce, the difference going to losses, inefficiencies and middlemen traders. Farmers in developed economies of Europe and the United States, in contrast, receive 64 to 81 percent of the price the local consumer pays for exactly the same produce in their supermarkets. Even though, India has shown remarkable progress in recent years and has attained self-sufficiency in food staples, the productivity of Indian farms for the same crop is very low compared to farms in Brazil, the United States, France and other nations. Indian wheat farms, for example, produce about a third of wheat per hectare per year in contrast with wheat farms in France. Similarly, at 44 million hectares, India had the largest farm area under rice production in 2009; yet, the rice farm productivity in India was less than half the rice farm productivity in China.

Other food staples productivity in India is similarly low, suggesting a major opportunity for growth and future agricultural prosperity potential in India. Indian total factor productivity growth remains below 2 percent per annum; in contrast, China has shown total factor productivity growths of about 6 percent per annum, even though China too has smallholding farmers. If India could adopt technologies and improve its infrastructure, several studies suggest India could eradicate hunger and malnutrition within India, and be a major source of food for the world.

Indian farms are not poor performing for every crop. For some, Indian farms post the best yields. For example, some of India's regions consistently posts some of the highest yields for sugarcane, cassava and tea crops every year.

Within India, average yields for various crops vary significantly between Indian states. Some Indian states produce two to three times more grains per acre of land than the grain produced in same acre of land in other Indian states. The table compares the statewide average yields for a few major agricultural crops within India, again for 2001-2002 agricultural year.

Crop **Kilogram per**	**Average farm Yield in Bihar Kilogram per Hectare**	**Average farm Yield in Karnataka Kilogram per Hectare**	**Average farm Yield in Punjab** **Hectare**
Wheat	2020	unknown	3880
Rice	1370	2380	3130
Pulses	610	470	820
Oil seeds	620	680	1200
Sugarcane	45510	79560	65300

Crop yields for some farms within India are within 90 per cent of the best achieved yields by farms in developed countries such as the United States and in European Union. No single state of India is best in every crop. Indian states such as Tamil Nadu achieve highest yields in rice and sugarcane, Haryana enjoys the highest yields in wheat and coarse grains, Karnataka does well in cotton, Bihar does well in pulses, while other states do well in horticulture, aquaculture, flower and fruit plantations. These differences in agricultural productivity within India is a function of local infrastructure, soil quality, micro-climates, local resources, farmer knowledge and innovations.

However, one of the serious problems in India is the lack of rural road network, storage, logistics network, and efficient retail to allow free flow of farm produce from most productive but distant Indian farms to Indian consumers. Indian retail system is highly inefficient. Movement of agricultural produce within India is heavily and overly regulated, with inter-state and even inter-district restrictions on marketing and movement of agricultural goods. The talented and efficient farms are currently unable to focus on the crops they can produce with high yields and at lowest costs.

One study suggests Indian agricultural policy should best focus on improving rural infrastructure primarily in form of irrigation and flood control infrastructure, knowledge transfer in forms of better yielding and more disease resistant seeds with the goal of sustainably producing as many kilograms of food staples per hectare as already produced sustainably in other nations. Additionally, cold storage, hygienic food packaging and efficient modern retail to reduce waste can also dramatically improve India's agricultural output availability and rural incomes.

The low productivity in India is a result of the following factors:

- The average size of land holdings is very small (less than 2 hectares) and is subject to fragmentation due to land ceiling acts, and in some cases, family disputes. Such small holdings are often over-manned, resulting in disguised unemployment and low productivity of labour. Some reports claim smallholder farming may not be cause of poor productivity, since the productivity is higher in China and many developing economies even though China smallholder farmers constitute over 97 percent of its farming population. Chinese smallholder farmer is able to rent his land to larger farmers, China's organised retail and extensive Chinese highways are able to provide the incentive and infrastructure necessary to its farmers for sharp increases in farm productivity.
- Adoption of modern agricultural practices and use of technology is inadequate, hampered by ignorance of such practices, high costs and impracticality in the case of small land holdings.
- According to the World Bank, Indian Branch: Priorities for Agriculture and Rural Development", India's large agricultural subsidies are hampering productivity-enhancing investment. Overregulation of agriculture has increased costs, price risks and uncertainty. Government intervenes in labour, land, and credit markets. India has inadequate infrastructure and services. World Bank also says that the allocation of water is inefficient, unsustainable and inequitable. The irrigation infrastructure is deteriorating. The overuse of water is currently being covered by over pumping aquifers, but as these are falling by foot of groundwater each year, this is a limited resource.

- Illiteracy, general socio-economic backwardness, slow progress in implementing land reforms and inadequate or inefficient finance and marketing services for farm produce.
- Inconsistent government policy. Agricultural subsidies and taxes often changed without notice for short term political ends.
- Irrigation facilities are inadequate, as revealed by the fact that only 52.6 per cent of the land was irrigated in 2003–04, which result in farmers still being dependent on rainfall, specifically the Monsoon season. A good monsoon results in a robust growth for the economy as a whole, while a poor monsoon leads to a sluggish growth. Farm credit is regulated by NABARD, which is the statutory apex agent for rural development in the subcontinent. At the same time overpumping made possible by subsidised electric power is leading to an alarming drop in aquifer levels.
- A third of all food that is produced rots due to inefficient supply chains and the use of the "Walmart model" to improve efficiency is blocked by laws against foreign investment in the retail sector.

INITIATIVES

The required level of investment for the development of marketing, storage and cold storage infrastructure is estimated to be huge. The government has not been able to implement various schemes to raise investment in marketing infrastructure. Among these schemes are *Construction of Rural Godowns*, *Market Research and Information Network*, and *Development/ Strengthening of Agricultural Marketing Infrastructure, Grading and Standardisation*. The Indian Agricultural Research Institute (IARI), established in 1905, was responsible for the search leading to the "Indian Green Revolution" of the 1970s. The Indian Council of Agricultural Research (ICAR) is the apex body in agriculture and related allied fields, including research and education. The Union Minister of Agriculture is the President of the ICAR. The Indian Agricultural Statistics Research Institute develops new techniques for the design of agricultural experiments, analyses data in agriculture, and specialises in statistical techniques for animal and plant breeding.

Recently Government of India has set up Farmers Commission to completely evaluate the agriculture programme. However the recommendations have had a mixed reception.

In November 2011, India announced major reforms in organised retail. These reforms would include logistics and retail of agricultural produce. The reform announcement led to major political controversy. The reforms were placed on hold by the Indian government in December 2011.

In the summer of 2012, the subsidised electricity for pumping, which has caused an alarming drop in aquifer levels, put additional strain on the country's electrical grid due to a 19 percent drop in monsoon rains, and may have helped contribute to a blackout across much of the country. In response the state of Bihar offered farmers over $100 million in subsidised diesel to operate their pumps.

THE CRISIS FACING INDIAN AGRICULTURE

The biggest problem Indian agriculture faces today and the number one cause of farmer suicides is debt. Forcing farmers into a debt trap are soaring input costs, the plummeting price of produce and a lack of proper credit facilities, which makes farmers turn to private moneylenders who charge exorbitant rates of interest.

In order to repay these debts, farmers borrow again and get caught in a debt trap. We will examine each one these causes which led to the current crisis in Andhra Pradesh, Kerala and Maharashtra, and analyse the role that liberalisation policies have played.

As was mentioned earlier, AP's experience is particularly relevant in this analysis because of its leadership. Let me explain in detail. Chandrababu Naidu, Chief Minister of Andhra Pradesh from 1995-2004, was an IT savvy neo-liberal, and believed that the way to lead Andhra Pradesh into the future was through technology and an IT revolution.

His zeal led to the first ever state level (as opposed to national level) agreement with the World Bank, which entailed a loan of USD 830 million (AUD 1 billion) in exchange to a series of reforms in AP's industry and government. Naidu envisaged corporate style agriculture in AP, and implemented World Bank liberalisation policies with great enthusiasm and gusto. He drew severe criticism from opponents, saying he was using AP as a laboratory for extreme neo-liberal experiments. Hence, AP's experience with liberalization is critical.

THE DEBT TRAP AND THE ROLE OF LIBERALISATION

The Debt Trap: High Input Costs

Seeds: The biggest input for farmers is seeds. Before liberalisation, farmers across the country had access to seeds from state government institutions. For example, AP's APSSDC3 produced its own seeds, was responsible for their quality and price, and had a statutory duty to ensure seeds were supplied to all regions in the state, no matter how remote. The seed market was well regulated, and this ensured quality in privately sold seeds too. (The damage done, 2005) With liberalization, India's seed market was opened up to global agribusinesses like Monsanto, Cargill and Syn Genta. Also, following the deregulation guidelines of the IMF, 14 of the 24 units of the APSSDC's seed processing units were closed down in 2003, with similar closures in other states.

This hit farmers doubly hard: in an unregulated market, seed prices shot up, and fake seeds made an appearance in a big way. Seed cost per acre in 1991 was Rs. 70 (AUD 2) but in 2005, after the dismantling of APSSDC and other similar organizations, the price jumped to Rs. 1000 (AUD 28), a hike of 1428%, with the cost of genetically modified pest resistant seeds like Monsanto's BT Cotton costing Rs. 3200 or more per acre, (AUD 91) a hike of 3555%. (Sainath, 2005) BT Cotton is cotton seed that is genetically modified to resist pests, the success of which is disputed: farmers in Andhra Pradesh and Maharashtra now claim that yields are far lower than promised by Monsanto, and there are fears that pests are developing resistance to the seeds. Expecting high yields, farmers invest heavily in such seeds. Also BT Cotton and other new seeds guarantee a much lower germination rate of 65% as opposed to a 90% rate of state certified seeds.

Hence 35% of the farmer's investment in seeds is a waste. (Sainath, 2004) Output is not commensurate with the heavy investment in the seeds, and farmers are pushed into debt. The abundant availability of spurious seeds is another problem which leads to crop failures. Either tempted by their lower price, or unable to discern the difference, farmers invest heavily in these seeds, and again, low output pushes them into debt. Earlier, farmers could save a part of the harvest and use the seeds for the next cultivation, but some genetically modified seeds, known as Terminator, prevent harvested seeds from germinating, hence forcing the farmers to invest in them every season.

Fertilizer and Pesticide: One measure of the liberalisation policy which had an immediate adverse effect on farmers was the devaluation of the Indian Rupee in 1991 by 25% (an explicit condition of the IMF loan). Indian crops became very cheap and attractive in the global market, and led to an export drive. Farmers were encouraged to shift from growing a mixture of traditional crops to export oriented 'cash crops' like chilli, cotton and tobacco. These need far more inputs of pesticide, fertilizer and water than traditional crops. Liberalisation policies reduced pesticide subsidy (another explicit condition of the IMF agreement) by two thirds by 2000. Farmers in Maharashtra who spent Rs. 90 an acre (AUD 2.5) now spend between Rs. 1000 and 3000 (AUD 28.5 – 85) representing a hike of 1000% to 3333%.

Fertilizer prices have increased 300% Electricity tariffs have also been increased: in Andhra Pradesh tariff was increased 5 times between 1998 and 2003. Pre-liberalisation, subsidised electricity was a success, allowing farmers to keep costs of production low. These costs increased dramatically when farmers turned to cultivation of cash crops, needing more water, hence more water pumps and higher consumption of electricity. Andhra Pradesh being traditionally drought prone worsened the situation.

This caused huge, unsustainable losses for the Andhra Pradesh State Electricity Board, which increased tariffs. (This was initiated by Chandrababu Naidu in partnership with Britain's DFID4 and the World Bank.) Also, the fact that only 39% of India's cultivable land is irrigated makes cultivation of cash crops largely unviable, but export oriented liberalisation policies and seed companies looking for profits continue to push farmers in that direction.

The Debt Trap: Low Price of Output

With a view to open India's markets, the liberalization reforms also withdrew tariffs and duties on imports, which protect and encourage domestic industry. By 2001, India completely removed restrictions on imports of almost 1,500 items including food. (The damage done, 2005) As a result, cheap imports flooded the market, pushing prices of crops like cotton and pepper down. Import tariffs on cotton now stand between 0 – 10%, encouraging imports into the country.

This excess supply of cotton in the market led cotton prices to crash more than 60% since 1995. As a result, most of the farmer suicides in Maharashtra were concentrated in the cotton belt till 2003 (after which paddy farmers followed the suicide trend). Similarly, Kerala, which is world renowned for pepper, has suffered as a result of 0% duty on imports of pepper from SAARC5 countries. Pepper, which sold at Rs. 27,000 a quintal (AUD 771) in 1998, crashed to Rs. 5000 (AUD 142) in 2004, a decline of 81%. As a result, Indian exports of pepper fell 31% in 2003 from the previous year. (Sainath, 2005) Combined with this, drought and crop failure has hit the pepper farmers of Kerala hard, and have forced them into a debt trap. Close to 50% of suicides among Kerala's farmers have been in pepper producing districts.

The Debt Trap: Lack of Credit Facilities and Dependence on Private Money Lenders

In 1969, major Indian banks were nationalized, and priority was given to agrarian credit which was hitherto severely neglected. However, with liberalisation, efficiency being of utmost importance, such lending was deemed as being low-profit and inefficient, and credit extended to farmers was reduced dramatically, falling to 10.3% in 2001 against a recommended target of 18%. A lack of rural infrastructure deters private banks from setting up rural branches, with the responsibility falling on the government, which has reduced rural spending as a result of its liberalisation policies.

Rural development expenditure, which averaged 14.5% of GDP during 1985 – 1990 was reduced to 8% by 1998, and further to 6% since then.

This at a time when agriculture was going through a crisis proved disastrous for farmers, who turned to private money lenders who charge exorbitant rates of interest, sometimes up to 24% a month. With input costs and output prices being what they are, coupled with crop failures and drought, they are pushed into debt which is impossible to repay. 12 out of India's 28 states have 50% and higher indebtedness among farm households. Andhra Pradesh has the highest percentage of indebted farm households — 82%. 64.4% of Kerala's farm households and 54.8% of Maharashtra's farm households are indebted (NSSO, 2003) Indebtedness has been identified as the single major cause of suicides in both Andhra Pradesh, Kerala and Maharashtra.

Liberalisation and How it Failed

Branco Milanovic, a World Bank economist describes how he believes liberalisation helps developing countries achieve growth: 'when a country lowers trade barriers, reduces government intervention in the market in order to allow market forces to operate freely, increases competition and attracts foreign investment, it will increase productivity and reduce inefficiency, which will lead to economic growth, and in a few generations, if not less, the poor will become rich, illiteracy will disappear, and poor countries will catch up with the rich.' This argument is an economic rationalist one, which views government intervention with profound suspicion, and has equally profound faith in unfettered market forces. What Mr. Milanovic neglects to mention, though, is that rich countries which now preach liberalisation protected their 'infant industries' at the time they began to industrialize, till they were strong enough to compete globally.

The US government, for example, had a protectionist trade policy in the late nineteenth century to help US companies become competitive in the world.

Besides, apart from wool, the US, Germany, Britain and France were all almost self-sufficient in the raw materials that they needed for industrialization, and took off from that platform, a luxury that India and other developing countries do not have. As German economist Friedrich List says, the adoption of these values (of liberalisation) assumes that all countries are at the same starting place, which (as we have seen above) is clearly not the case. In fact, it is this very reason that has brought about the crisis that Indian agriculture is facing today.

Most farmers in India were already in a position of minimum security, with no education system, credit facilities, access to alternative employment, or efficient technology. Their only support was government subsidy and regulation. Liberalisation policies came in and dismantled their only support structure. It halted the sharp reduction in rural poverty from 55% in the 1970s to 34% in the 1980s. Not only has the incidence of poverty in rural areas not gone lower than 34% in the 1990s, it has gone to higher levels of 42% in individual years.

The second most popular argument of the economic rationalists in favour of liberalisation is that competition will weed out the inefficient, and in the growth that ensues, employment will be provided in other areas of the economy, thus lifting the poor out of poverty. This argument however assumes that the poor will be able to take advantage of the opportunities presented to them. As Robert Issac says in 'The Globalization Gap', "Globalization encourages the well positioned to use tools of economics and politics to exploit market opportunities, boost technical productivity, and maximize short-term material interests." This is compounded in India, where the gap between one who is 'well positioned' and one who is not can be extreme. With a lack of investment, chances of generation of rural employment are slim.

Unemployment and underemployment are chronic problems in India, with the rate of unemployment being close to 10% in 2004. Primary education in rural areas is mismanaged and bad quality, and there is no system which helps agricultural workers find alternate employment, or develop alternate skills. In the face of such obstacles, it is nearly impossible to expect agricultural workers to shift to alternate fields. Coming back to AP, the IT Revolution spearheaded by Chandrababu Naidu attracted companies like Google, Amazon, Microsoft and Dell and created thousands of jobs. However, given the skills and education of most farmers, it is obvious that none of this translated into job opportunities for them.

The final argument that supporters of globalization have is the much touted 10% reduction in poverty (60 million decline in poor) in India in the year 2000. However, this figure was challenged by experts. Poverty is defined according to how many people consume less than the nutritional minimum prescribed. (2400 calories for rural areas, and 2100 for urban areas) Major changes in survey design in 1999-2000 not only made the resultant estimates incomparable to previous years' estimates, but an over-estimation of consumption (meaning people were getting enough food, hence were not considered poor) meant a sharp reduction in poverty figures. After experts challenged it, the Planning Commission of India accepted that the figure was inaccurate, and could not be compared to previous years' estimates, hence the 10% drop in poverty is incorrect. With adjusted figures, experts have determined that the decrease in poverty was a mere 2.3%, and that the number of poor increased by nine million in 2002 as compared to 1999.

Liberalization and 'Growth'

Many economists now concede that the relationship between liberalisation and growth are 'uncertain at best'. According to the Centre for Economic and Policy research, which studied impact of liberalisation reforms on the developing world, key economic and social indicators such as increases in life expectancy, infant and child mortality, education and literacy levels slowed down in the 20 years between 1980 and 2000 when liberalisation policies were implemented, compared to the 20 years leading to 1980. (The damage done, 2005) This defeats the economic rationalist argument of free trade eliminating poverty, since the 20 years leading up to 1980 witnessed high protectionist policies and trade barriers. Following the suicides in 2000, the World Bank and Britain's DFID abandoned power reforms in Andhra Pradesh four years before schedule. It admitted that it had 'substantially underestimated' the 'complexity of the process' and that there must be 'increased consultation with the farmers to get their acceptance' of any further reform.

The Andhra Pradesh government sponsored report by the Commission of Farmer's Welfare squarely laid the blame for its agrarian crisis on the state and central government's policies: "While the causes of this crisis are complex and manifold, they are they are dominantly related to public policy. The economic strategy of the past decade at both central government and state government levels has systematically reduced the protection afforded to farmers and exposed them to market volatility and private profiteering without adequate regulation; has reduced critical forms of public expenditure; has destroyed important public institutions, and has not adequately generated other non-agricultural economic activities." A report on suicides in Kerala similarly held the liberalization policies of the government responsible.

DIVERSIFICATION TRENDS OF INDIAN AGRICULTURE

Farming is the backbone of Indian economy. The sector plays a vital role in the development of India with over 60 per cent of the country's population deriving their subsistence from it. Most of the industries also depend upon the farming sector for their raw materials.

India ranks first in the production of milk, pulses, jute and jute-like fibres; second in rice, wheat, sugarcane, groundnut, vegetables, fruits and cotton production; and is a leading producer of spices and plantation harvests as well as livestock, fisheries and poultry. The rapid growth of farming is essential not only for self-reliance but also for meeting the food and nutritional security of the people, to bring about equitable distribution of earnings and wealth in rural areas as well as to reduce poverty and improve the quality of life.

Department of Agriculture and Cooperation under the Ministry of Agriculture is the nodal organisation responsible for the development of the farming sector in India. The organisation is responsible for formulation and implementation of national policies and programmes aimed at achieving rapid farm growth through optimum utilisation of land, water, soil and plant resources of the country.

MARKET DYNAMICS

India has improved its position in the farm and food exports to 10th globally, backed by policy impetus by the government. "Exports of farm products are expected to cross US$ 22 billion mark by 2014 and account for 5 per cent of the world's farming exports," according to the Farm and Processed Food Products Export Development Authority (APEDA).

Total exports of Indian Agri and processed food products from April 2012 to March 2013 stood at ₹ 11,633,168.41 lakh (US$ 17.26 billion) as compared to

₹ 8,248,025.32 lakh (US$ 12.23 billion) during the same period last year, according to the data provided by APEDA.

India recorded an increase of 22 per cent in the export of spices and spice-based products during 2012-13 to touch 699,170 tonnes, as against 575,270 tonnes in the previous financial year. The Soymeal exports during June 2013 was 213,400 tonnes as compared to 180,900 tonnes in the same period of previous year, registering an increase of 18 per cent.

Groundnut sowing in Gujarat has touched 1.41 million hectares as on July 1, 2013 as against 0.22 million hectares in the corresponding period last year, according to Agriculture andthe Co-operation Department, Government of Gujarat.

The foreign direct investment (FDI) inflows in farm services and machinery sector during April 2000 to June 2013 stood at US$ 1,620.65 million and US$ 337.21 million respectively, as per the data released by the Department of Industrial Policy and Promotion (DIPP).

MAJOR DEVELOPMENTS AND INVESTMENTS

A number of memorandum of understandings (MoU) was signed between the Indian Council of Agricultural Research (ICAR) and the industry. MoUs have been signed for more than 60 ready-to-commercialise agro-technologies from different farm sectors like harvests, horticulture, food technology, veterinary, Agri-engineering, Agri-inputs and fisheries.

The Abu Dhabi-based Al Dahra International Investment LLC is set to invest about ₹ 112 crore (US$ 16.61 million) in Kohinoor Foods Ltd (KFL).

Private equity (PE) and venture capital (VC) firms invested US$ 126 million across nine Indian agribusiness companies during the first six months of 2013, as per data from Venture Intelligence.

Coromandel International Ltd, India's leading fertilizer manufacturer, with its joint venture (JV) partners have inaugurated a 1.4 million tonne (MT) phosphoric acid plant in Tunisia. The Indo-Dutch joint initiative in farming envisages setting up about 10 centers of excellence (CoE) in Punjab, Gujarat, Kerala, Maharashtra and Karnataka in the next few years, a move that could help raise output and yields.

The total outlay of ₹ 27,049 crore (US$ 4.01 billion) has proposed to the Ministry of Agriculture in the Union Budget 2013-14, which is 22 per cent more than the revised estimates by the year 2012-13. Further, ₹ 1,000 crore (US$ 148.38 million) has been allocated to continue support to the new green revolution in Eastern States like Assam, Bihar, Chhattisgarh and West Bengal to increase the rice production. An outlay of ₹ 500 crore (US$ 74.19 million) is also proposed for starting a programme of harvest diversification that would promote scientific innovation and encourage farmers to choose harvest alternatives in the original green revolution States.

GOVERNMENT INITIATIVES

Some of the major initiatives taken by the Government of India are:

- The Government of India has set a target of 259 million tonnes (MT) of foodgrains production in 2013-14. It is implementing various harvest development programmes/ schemes for achieving production targets of various harvests
- The government has permitted 100 per cent FDI under the automatic route, subject to certain conditions in Floriculture, Horticulture, Apiculture and Cultivation of Vegetables and Mushrooms under controlled conditions; Development and production of Seeds and planting material; Animal Husbandry (including breeding of dogs), Pisciculture, Aquaculture, under controlled conditions; and Services related to agro and allied sectors. 100 per cent FDI is also permitted in the tea sector.
- For ensuring quality of Agrochemicals, the government has set up 71 pesticides testing laboratories across the country that include 68 state laboratories, 2 regional laboratories and 1 central laboratory
- The Government of Punjab gave its approval to the expansion plan of state-owned Milkfed entailing setting up of four mega milk plants worth ₹ 250 crore (US$ 37.09 million) each in the state
- The government has launched an SMS Portal for Farmers on July 16, 2013 for disseminating relevant information, giving topical and seasonal advisories and providing services through SMSs to farmers in the language of the State

  Under the Union Budget 2013-14,
- The government has substantially improved the availability of farm credit to improve investments in the farm sector. The annual farming credit target for the financial year 2013-14 has increased to ₹ 7 lakh crore (US$ 105.96 billion) from ₹ 5.75 lakh crore (US$ 85.32 billion) in 2012-13
- The allocation for the Integrated Watershed Programme has been increased to ₹ 5,387 crore (US$ 799.37 million) from ₹ 3,050 crore (US$ 452.58 million) to provide relief to small and marginal farmers especially in drought prone and ecologically-stressed regions
- The National Livestock Mission will be launched to attract investment and to enhance productivity of livestock, taking into account local agro-climatic conditions. ₹ 307 crore (US$ 45.55 million) have been provided to the Mission

ROAD AHEAD

With nearly a 1.2 billion population, India requires a robust, modernisedfarming sector to ensure the food security for its population.

In order to meet the food grain requirements of the country, the farm productivity and its growth need to be sustained and further improved. The growth target for farming in the 12th Five Year Plan is estimated to be 4 per cent as compared to 3.6 percent for the 11th Plan.

The Ministry of Agriculture is promoting a new strategy for farm mechanisation through its various schemes and programmes. A dedicated Submission on Farm Mechanisation has been proposed in the 12th Plan which includes custom-hiring facilities for farm machinery as one of its major components.

DIVERSIFICATION TRENDS OF INDIAN AGRICULTURE

DIVERSIFICATION EXPLAINED

Diversification of farming refers to the shift from the regional dominance of one harvest to regional production of a number of harvests, to meet ever-increasing demand for cereals, pulses, vegetables, fruits, oilseeds, fibres, fodder and grasses, fuel, etc. It aims to improve soil health and a dynamic equilibrium of the Agri-ecomethod. Harvest diversification takes into account the economic returns from different value-added harvests.

It is different from the concept of multiple harvesting or succession planting in which multiple harvests are planted in succession over the course of a growing season. Moreover, it implies the use of environmental and human resources to grow a mix of harvests with complementary marketing opportunities, and it implies a shifting of resources from low value harvests to high value harvests, usually intended for human consumption such as fresh market fruits and vegetables. With globalisation of the market, harvest diversification in farming means to increase the total harvest productivity in terms of quality, quantity and monetary value under specific, diverse agro-climatic situations world-wide.

There are two approaches to harvest diversification in farming. First is horizontal diversification, which is the primary approach to harvest diversification in production farming. Here, diversification takes place through harvest intensification by adding new high-value harvests two existing harvestingmethods as a way to improve the overall productivity of a farm or region's farming economy.

The second is the vertical diversification approach in which farmers and others add value to products through processing, regional branding, packaging, merchandising, or other efforts to enhance the product. Opportunities for harvest diversification vary depending on the risks, opportunities and the feasibility of proposed changes within a socioeconomic and agro-economic context. Harvest diversification may occur because of government policies. The "Technology Mission on Oilseeds", "Spices Development Board", "Coconut

Development Board" etc. Are examples where the Indian government created policies to thrust change upon farmers and the food supply chain at large as a way to promote harvest diversity.

Harvest diversification is the outcome of several interactive effects of many factors:

1. Environmental factors including irrigation, rainfall, temperature, and soil fertility.
2. Technology-related factors including seeds, fertilizers and water technologies, but also those related to marketing, harvest, storage, agro-processing, distribution, logistics, etc.
3. Household-related factors including regional food traditions, fodder and fuel as well as the labour and investment capacity of farm people and their communities.
4. Price-related factors including output and input prices as well as national and international trade policies and other economic policies that affect the prices either directly or indirectly.
5. Institutional and Infrastructure-related factors including farm size, location and tenancy arrangements, research, in-field technical support, marketing methods and government regulating policies, etc.

All these five factors are interrelated. The adoption of harvest technologies is commonly assumed to be influenced primarily by resource-related factors when institutional and infrastructure factors can play as much or more of a role in their adoption.

AREA EXPANSION PROBLEMS UNDER RICE AND WHEAT HARVESTS

Scaling up production area poses several new problems of significance such as:

- Excessive use of groundwater leading to poor water use efficiency and depletion of groundwater.
- Deterioration of soil health or soil fertility.
- Multiple infestations of weed flora, insect pests and diseases.
- Indiscriminate use of energy such as chemical, electrical or disease, etc.
- Reduction in the availability of other protective food and high value harvests.
- Pollution of agro-Eco methods.

On the other hand, harvest diversification has potential as an economic driver in farming regions. It may prove to be of paramount importance in meeting challenges that arise from a post-green revolution scenario.In view of shrinkage of farm land and operational holdings due to the expansion of urban centers, changes in consumer food habits, exponential population growth rate, farmers are pressured to include or substitute additional harvests in to the harvestingmethod.

MAJOR DRIVING FORCES FOR HARVEST DIVERSIFICATION

The major driving forces for harvest diversification are:

- Increasing earnings on small farm holdings.
- Withstanding price fluctuation.
- Mitigating ill-effects of aberrant weather.
- Balancing food demand.
- Improving fodder for livestock animals.
- Conservation of natural resources (soil, water, etc.).
- Minimizing environmental pollution.
- Reducing dependence on off-farm inputs.
- Decreasing insect pests, diseases and weed problems.
- Increasing community Food Security

Indian farming is characterised by a dominance of small and marginal farmers (almost 68 per cent) who suffer as a result of difficult socioeconomic conditions. 75 per cent of the farm holdings are below 2 hectares, and a large portion of rural people subsists as small holders.

Earnings from these farmscannot be raised up to the desired level to sufficiently alleviate poverty in the countryside unless existing harvest production methods are diversified through inclusion of high value horticultural and arable harvests. Furthermore, increased dependence on one or two major cereal harvests (wheat, rice, etc.) Witnessed after the green revolution makes the farming economy vulnerable to price fluctuation rising due to demand-supply or export-import equations especially after the WTO began influencing markets.

Harvest diversification on the other hand, can better tolerate the ups and downs in the market value of farm products and may ensure economic stability for the farm families of the country. The adverse effects of aberrant weather, such as erratic and insufficient rainfall and drought are very common in a vast area in farm production of the country. Incidence of flood in one part of the country and drought in the other part is a very frequent phenomenon in India. Under these aberrant weather situations, dependence on one or two major cereals (rice, wheat, etc.) Is always risky.

Hence, harvest diversification through substitution of one harvest or mixed harvesting/inter-harvesting may be a useful tool to mitigate problems associated with aberrant weather to some extent, especially in the arid and semi-arid drought-prone/dryland areas.

IMMEDIATE NEED

Harvest diversification in farming in India is taking place vertically or horizontally, mostly due to market forces and occasionally due to domestic needs.

Where there are concerns regarding land and water use and quality, there is an immediate need to consider:

- There is a need to generate place-based approaches for diverse farming situations under various socioeconomic conditions, domestic needs, market infrastructure, input supply, etc.
- A concept of sustainable productivity for each unit of land and water through harvest diversification needs to be fostered.
- Processing of farm produces into value added products offered scope for employment in non-farm works such as distillation of active ingredients from medicinal and aromatic plants (herbal products), scope of industrialisation in farming for sugar, paper board manufacturing, etc. To increase employment in rural areas.
- The research on harvest diversification is best done in a farmer-participatory mode in which a multi-disciplinary team of scientists involves farmers from project planning through arriving at conclusions.
- There is a need for promoting co-operatives in rural areas to solve micro-level and location-specific problems.
- Major thrust should be given on horticulture (vegetables, fruits, flowers, spices, etc.) And animal husbandry (dairying, poultry, jittery, piggery, duckery, etc.) To support a vigorous and expanding export market, balanced with supplying local markets with affordable, healthy food.
- Strengthening food processing and other value-added industries in rural areas is a means to provide employment to rural youth.
- There is a need to develop rural infrastructure such as roads, markets, medical and educational facilities in the villages with efficient utilisation of local resources for farming community in a more pragmatic way.
- Harvest diversification provides efficient use of farm inputs and contributes to a strong rural economy.
- Alternate harvestingmethods and farm enterprise diversification are most important for generating higher earnings, employment and protecting the environment.
- There are numerous opportunities to adopt subsidiary occupations to the rice-wheat harvestingmethods common in India. These include vegetable cultivation, fruit cultivation, floriculture, medicinal and aromatic plant cultivation, mushroom cultivation, dairying, piggery, goatery, poultry and Duckery, fishery or aquaculture, beekeeping, Agroforestry, biodiesel farming with Jatropha Curcas (veranda), palm, neem, Karanja, etc. To provide ample scope for diversification of rice-wheat harvestingmethod in northwestern and south India and north-eastern states.

- Enterprise diversification generates more earnings and rural employment year round.

CONCLUSION

Diversification in farming' has tremendous impact on the agro-socio-economic impact and uplifting of resource-poor communities. It generates earnings and employment for rural youth year round for the ultimate benefits of the farmers in the country. It implies the use of local resources in a larger mix of diverse harvestingmethods and livestock, aquaculture and other non-farm sectors in the rural areas. With the globalisation of markets in the WTO era, diversification in farming is one means to increase the total production and productivity in terms of quality, quantity and monetary gains under diverse agro-climatic situations of the country. There are many opportunities of harvest diversification both in the irrigated and non-irrigated vast areas in the rural India.

4

Colonial Aspects of Exploitation in the Indian Countryside

The imposition of Colonial Rule in India led to a drastic break with the past, in that not only did the scale and intensity of the exploitation of the village communities greatly increase, it also led to the introduction of new, and almost entirely parasitic intermediaries between the state and the tax paying masses - intermediaries who were typically induced (or pressurized) to abandon traditional restraint, and discard the old formulae that helped mediate the burden on the typical village peasant or artisan.

(It is notable that whereas ancient texts such as the Manusmriti put a cap on agricultural taxes at one-sixth of the produce, and taxes during Mughal rule had risen to something between a third and almost half the produce, during British rule, the peasant was often left with barely a third of the crop.)

Not only were these intermediaries under no pressure to build or maintain traditional water-management structures, the manner in which they were chosen by the British to collect the taxes actually discouraged any diversion of the taxes towards any capital upkeep. Over time, this not only led to the destruction of century-old water management structures, but also to the virtual destruction of the knowledge systems and cultural traditions that had helped build and preserve these water-management techniques over the centuries in states such as Bihar, Bengal, Karnataka, Tamil Nadu and others.

In addition, the British obliged these intermediaries to collect the taxes not in kind, but in cash. Since the peasant had little experience or understanding of a cash economy (or any control on the price of grain and agro-inputs), this put a further burden on the peasant, who now also had to face the thuggery of unscrupulous grain traders and usurious money-lenders - who took full advantage of the highly diminished status of the Indian peasant in the colonial dispensation.

(Note that the existence of money-lenders and traders is documented from at least Ashokan times, but the traditional Indian state typically insisted on a cap on interest payments and profits. Moreover, money-lenders and traders were traditionally active mostly in towns, not in villages. The encroachment of the cash economy and the growth of usury in the countryside began to adversely affect the peasants during the later Mughal period, but the problem grew very dramatically during colonial rule.)

As a result, not only was the typical Indian peasant reduced to a state of utter degradation, social relations within the village also became highly distorted. The traditional solidarity that had existed between villagers was now subject to the divisive and ruinous tactics of the parasitic intermediaries who had the protection and support of the colonial state.

As colonial rule progressed, the typical Indian peasant (and village artisan) faced a dual burden. Not only did the traditional feudal form of exploitation greatly intensify (in comparison to traditional Indian practices), the typical Indian villager was now also subject to the merciless forces of mercantile capitalism, who attempted to extract every ounce of savings or assets that any villager might still possess. Eventually, the typical Indian villager was stripped of all savings, and driven to mortgaging a considerable portion of any assets - whether personal jewelry, land and livestock, or tools and equipment.

But even as the Indian money-lenders and grain traders fleeced the village artisans and peasants, their accumulated wealth was not especially significant in comparison to the British mercantile capitalists of the East Indian Company who kept a lion's share of the loot.

It is especially important to note that whereas a share of the village surplus went to a few Indian intermediaries (who served as a buffer between the British colonial over-lords and the Indian masses, and generally acted to quash the class struggle), it was the British who expropriated the bulk of the peasant's assets and annual agro-produce, and transferred most of this looted wealth to the British homeland where it was turned into industrial capital and urban development.

This point is of special significance in understanding the unique nature of class contradictions in the Indian countryside (as compared to elsewhere in the non-colonized world.) For instance, almost throughout Europe, the peasantry was also subject to exploitation by the feudal landlords, but the expropriated surplus from the countryside eventually found its way into the urban banking system, and ultimately facilitated in the growth of industrial capitalism. Sections of the impoverished peasantry then had the chance to gradually move into industrial occupations.

But India was subject to an especially tragic plight. Its masses were thoroughly impoverished, but their immediate exploiters were able to keep only a very small share of what had been extracted from the peasantry. This meant that even when a transformation into industrial capitalism became possible (after freedom from British colonial rule) - there was simply not enough accumulated savings to fund the rapid expansion of industry that could eventually transform the countryside. When the British left, they left India with extremely limited material means to industrialize, but a sizeable class of parasitic intermediaries who would serve as a constant drag on future development. Unlike India's pre-colonial mercantile classes and the more nationalist sections of India's traditional feudal nobles and intermediaries (who were mostly neutralized or exterminated in 1857) this new colonial era mercantile class was extremely backward in cultural terms and was generally lacking in nationalist spirit.

In comparison, the mercantile classes in Europe of the late 18th and 19th centuries were not only wealthier, but also more urbane, and displayed a much higher level of cultural, scientific and technological openness and curiosity. This was also generally true of Japan, and to a somewhat lesser degree, of mercantile classes in other parts of Asia that escaped direct European colonial rule (such as in Korea, Thailand and even China).

It should also be noted that whereas the mercantile classes in Europe, Japan, Korea and China were moving in a broadly secular direction, the Indian mercantile classes were excessively weighed down by sentiments favoring religious revivalism or extreme sectarianism (a trend that was especially strong amongst Muslims, but also posed a serious problem amongst Sikhs and Hindus).

It is thus important to understand that not only was the Indian countryside weighed down by a new and more intensified feudal oppression in the colonial era (such as through the Zamindari or Mahalwari systems), it was also oppressed by a highly reactionary and socially conservative intermediary class of petty mercantile capitalists who would pose as the biggest obstacle to any radical or revolutionary developments in the Indian countryside.

Finally, it should be noted that colonial rule exacerbated the problems of the Indian countryside in yet another way. Prior to colonization, India was steadily becoming more urbanized, with some estimates suggesting that as much as a third of the Indian population may have been living in large or small towns.

In addition, even in the villages, a considerable proportion of the population comprised of skilled artisans - weavers, potters, carpenters, metal-workers, painters etc. The proportion of the population that was exclusively engaged in agricultural production was steadily decreasing.

But in the colonial regime, several laws were passed that led to a catastrophic de-urbanization and de-industrialization of India. Trade tariffs and excise duties were set so as to destroy Indian industries, and squeeze domestic trade. In states like Bihar and Bengal, severe restrictions were placed on the use of inland water-ways - causing fishing and inland shipping and transportation to suffer. This led to even greater pressures on agriculture since large categories of highly skilled artisans and non-agricultural workers were thrown out of work.

Effects of Partition

The problem for India was compounded further by the demands for partition, and the insistence of the Muslim League in splitting the country on communal lines. It is important to note that prior to partition, the best naturally irrigated and most productive lands in Northern India were in North Western Punjab (now Pakistan). Fed by a network of all-weather rivers, no other part of the Indian sub-continent was capable of producing more grain with relatively little capital investment. Similarly, Bangladesh was a much better irrigated land than West Bengal (although it faced far greater risks of flooding than Pakistan Punjab).

But in any case, it is worth pointing out that the bulk of pre-partition India's wheat, cotton and jute were grown in what are now Pakistan and Bangladesh. Not only did partition provide India with a smaller share of the land (relative to the ratios of the Hindu, Sikh and Muslim populations), it also left India without the means to feed itself, and without some of the essential raw materials for industry. (Although India did inherit bulk of the cotton and jute mills, India was unable to take full advantage of this until it had increased its own production of jute and cotton to supply these mills, whereas the raw material producers in West and East Pakistan could at least export their raw materials)

It should also be noted that partition disrupted what had been a somewhat natural distribution of labor between what became Pakistan and India. Since West Punjab and Sindh were better suited to agriculture, the relatively arid states of Rajasthan and Gujarat specialized in all manner of cottage industries and manufacturing that needed the raw materials and markets of West Punjab and Sindh to flourish.

In the same way, there was considerable co-dependencies not only between Western and Eastern Bengal, but also between undivided Bengal and its neighboring states such as Assam, Orissa and Bihar. The artificial vivisection of Bengal virtually destroyed the potential of regenerating the traditional trade linkages that had been so vital to the prosperity of the Eastern half of India. Geographically, it should be evident that the management of water and other natural resources cannot be efficiently undertaken without the effective coordination and active involvement of all the states in the Ganga-Brahmaputra region.

But the vitiated atmosphere under which partition took place meant that the possibilities of mutually beneficial cooperation in matters of resource-sharing, trade, transportation and agriculture were practically foreclosed. (To this day, communal forces in Bangladesh have resisted attempts by India to expand cooperation since they would rather see India suffer more rather than see both India and Bangladesh make more rapid progress.)

Thus, when the British left, India had become a much more village-based agricultural economy than it had been earlier - it is likely that the British had pushed India back by several centuries in this respect. Yet, its prospects for rapid industrialization were also quite bleak.

India had inherited one of the most depressing scenarios in Asia - a fairly densely populated nation with a pitiful urban base, a rural infrastructure in wrecks, a huge mass of population forced to survive exclusively on agricultural production - yet cynically oppressed by petty mercantile capitalists and feudal intermediaries. Moreover, it had also been left artificially divided on communal lines that meant that important historical linkages might never be revived (or revived only with great difficulty). No civilized nation could have faced a worse set of circumstances.

THE INDIAN POLITICAL SITUATION POST-1947

The Indian National Congress conducted a dual (and somewhat contradictory) policy relative to Indian agriculture. On the one hand, it attempted to ameliorate the problems of the Indian countryside by investing in irrigation, agro-research, agro-modernization and mechanization, but on the other hand, it was only half-hearted (at best) in implementing its own platform of land reforms because it was reluctant to hurt the interests of the feudal and mercantile intermediaries that oppressed most peasants.

Where the struggle for 'land to the tiller' became very intense, it was obliged to make concessions to the impoverished peasants, and move against the parasitic forces, but wherever the demands for land reform were not backed up by strong and effective people's movements, it chose to drag its feet.

It is only in the sixties, when India's Naxal movement began to spread and acquire a certain degree of political depth did the Indira Congress feel the threat and necessity of clipping the wings of the Zamindars, taking away the privy purses of the Maharajas, and assisting farmers more actively. Even as the Naxal movement was harshly repressed, banks were nationalized and were required to set up more branches in rural areas so that loans for seeds and farm equipment (and inputs such as fertilizers and pesticides) could be purchased through loans at more reasonable interest rates. The state also actively intervened in the buying and selling of grain.

Although these measures did not help all sections of the peasantry, it did allow a section of the country's farmers to gradually liberate themselves from the clutches of the old parasitic intermediaries, and move towards mechanized farming. Over time, peasants struggles put further pressure on the government leading to periodic waiving of loans, and easing of credit norms so that even middle farmers could get loans from government banks and rural credit unions. Subsidies were also provided for electricity and water use, as well as on the sale price of fertilizers and pesticides.

Thus, a combination of people's pressure and the gradual expansion of industrial capitalism has brought about significant changes in the Indian countryside. Whereas a section of Indian farmers (at least in some states) have become rich capitalist farmers, a large middle layer of farmers has also developed which is able to survive (even eke out a small surplus) in the good agricultural years (when the monsoon and market conditions happen to be favorable). These middle-income farmers also hire seasonal labor - and this further distinguishes them from the poorest farmers who can't even survive in the good years. Thus today, there is a growing differentiation in the Indian countryside, and this is not too different from what has been seen in many other relatively mature capitalist countries. However, the increasing sub-division of land, and the rising costs of agricultural inputs is a constant source of problems and tensions in the countryside.

Comparisons with Europe

In many respects, developments in Indian agriculture have paralleled developments in the Germany and Italy in an earlier century. In both Germany and Italy, the transformation towards a modern industrial society took place slowly, and through a series of evolutionary steps - as is occurring in India today.

The main differences between India and Germany and Italy are simply that not only did Germany and Italy escape the horrors of colonization, they indirectly benefited from the colonial loot. This meant that as their urban industrial economies expanded, these nations were able to absorb their landless populations in their rapidly expanding industrial and services sectors.

When industrial expansion wasn't fast enough, Europe's landless had another escape route - external migration. For more than a century, Europe's unemployed and poor escaped to the US, Canada, South America, South Africa, Australia and New Zealand - where the indigenous populations were killed (or displaced from the best agricultural lands). Whereas Europe had the distinct advantage of being able to export away its rural poor to land it grabbed from indigenous communities (who were not well-armed or skilled in modern warfare), India cannot simply wish away its excess rural population.

Solutions to the Problems of Indian Agriculture

The subdivision of already small farms in an agricultural model whose profitability is greatly linked to economies of scale makes the survival of the small farmer very difficult. Although the introduction of tractors and other industrial implements, of chemical fertilizers and pesticides, greatly increased the productivity of land and labor, it has not come without considerable burdens for the small farmer who is finding it very hard to compete with large and medium farmers. In most industrial nations, the introduction of such practices in the countryside has been accompanied by a gradual reduction in the rural population so that those who remain on the land can enjoy so.

AGRICULTURAL POLICY AND LONG-TERM AGRICULTURAL GROWTH

Agriculture plays important roles in economic development, such as provision of food to the nation, enlarging exports, transfer of manpower to nonagricultural sectors, contribution to capital formation, and securing markets for industrialization. Improvement in agricultural productivity is key to the realization of each of these roles and herein lies the importance of agricultural policies.

Historical records have shown that agricultural productivity has been growing due to introduction of modern technologies, commercialization of agriculture, capital deepening, factor shifts from agriculture to nonagricultural sectors, etc., none of which occurred in isolation from the policy settings of each agriculture. This chapter, therefore, attempts to associate long-term trends in agricultural productivity in India and Pakistan with changes in political institutions and agricultural policies. The use of a very long time series that corresponds to the current border of India and Pakistan distinguishes this chapter from the existing studies on South Asian agriculture.

At the same time, when discussing the effects of policies, the fictitious treatment of "India" and "Pakistan" that refer to the areas delineated by the current international border becomes more problematic. Because of this reason, we use the pre-1947 performance as a benchmark to normalize the post-1947 performance and discuss the policy impacts during the post-1947 period only.

DATA

This chapter employs an updated version of the dataset compiled by the author. We cover a period from 1901/02 to 2001/02. The addition of the 1990s distinguishes this chapter from the previous studies. We cover geographic areas that correspond to the current international borders of India and Pakistan. The data compilation procedures for the colonial period is explained by Kurosaki (1999).

We basically subtracted areas and production corresponding to the areas currently in Pakistan and Bangladesh from the data base compiled by Sivasubramonian (1960, 1997). Information included in district-level data in *Season and Crop Reports* from Punjab, Sind (or Bombay-Sind), the NWFP (North-West Frontier Province), and Bengal was utilized in this exercise. Because historical data on agriculture are less detailed than the current ones, we cannot compile production statistics for several crops.

Therefore, we focus on production of principal crops, which are important in contemporary India and Pakistan, and for which detailed data on production and prices are available from the British period. For India, eighteen crops are included: foodgrains = rice, wheat, barley, jowar (sorghum), bajra (pearl millet), maize, ragi (finger millet), gram (chickpea); oilseeds = linseed, sesamum, rapeseed & mustard, groundnut; other crops = sugarcane, tea, coffee, tobacco, cotton, and jute & mesta.

These crops currently occupy more than two thirds of the total output value from the crop sector and more than a half of the total output from agriculture, and their contribution was higher in the colonial period. For Pakistan, all twelve crops included in the major crops subsector are covered, *i.e.*, rice, wheat, barley, jowar, bajra, maize, gram [foodgrains]; rapeseed & mustard, sesamum, sugarcane, tobacco, and cotton. These crops currently occupy about 70 percent of value-added from all the crops and about 40 percent of value-added from agriculture, and its share was higher in the colonial period.

The gross output values from these crops are aggregated using 1960 prices. Ideally, the sum of value-added evaluated at current prices and then deflated using some price index would be a better measure, but the sum of gross output values at constant prices is used as a proxy due to the absence of reliable data on input prices and quantities before independence. The results reported in this chapter are insensitive to the choice of other base years (1938/39 and 1980/81). We call the gross output value thus compiled Q. As measures for partial productivity, we divide Q by L (official population estimates of India and Pakistan) or by A (the sum of the acreage under the eighteen or twelve crops).

PRODUCTIVITY AND CROP MIX IN AGRICULTURE IN INDIA AND PAKISTAN

We updated tables and figures reported in Kurosaki (1999, 2002). Since the data revisions in the period already covered in these studies are minor, the discussion in this chapter focuses on the latest decade. Namely, we investigate whether or not trends in agricultural productivity and crop mix have changed during the 1990s.

Productivity

Growth performance in Indian agriculture is plotted. The statistical results to quantify the level of growth rate, its significance, and its variability in India. To smooth out short-term fluctuations, we estimate a time series model for Yt in the form of

(1) $\ln Yt = a0 + a1\, t + ut,$

where t is measured in years and ut is an i.i.d. error term. We estimate equation (1) for the logarithm of Q, Q/L, and Q/A, by the OLS for each decade and then for the periods before and after independence. The first column of Table below gives coefficient estimates for $a1$.

The total output value (Q) grew very little up to the Partition and then grew steadily since then. The growth rate in the 1990s was 1.7%, which is less than the post-Partition average of 2.8%. The column for "Variability" shows how variable was the output around the fitted values in terms of the coefficient of variation. The 1990s was associated with less variability. In the 1990s, output per capita (*Q/L*) did not grow while output per acre (*Q/A*) grew at a rate similar to that of *Q*. In other words, the source of growth in the 1990s in India was exclusively the improvement in aggregate land productivity.

Similar changes were observed in Pakistan but with higher growth rates throughout the period. The growth rate of *Q* before the Partition was smaller than the post-Partition period but still statistically significant at 1.3%. The growth rate declined in the 1990s to 2.3% against the post-independence average of 3.6%, but its level was still higher than that of India. Unlike in India, the variability of *Q* was similar to the previous periods and the source of growth was not only the improvement in aggregate land productivity but also the expansion of total acreage in Pakistan. The growth rate of output per acre (*Q/A*) was positive and significant but smaller than that of *Q*. in the 1990s in Pakistan.

Table. Growth Performance of Indian Agriculture

	Q (Total output value)		*Q/L (Output per capita)*
CQ/A (Output per acre)			
Period	*Growth rate*	*Coeff. var.*	*Growth rate*
Coeff. var.	*Growth rate*	*Coeff. var.*	
1901/02 – 1910/11	1.04%	11.8%	0.39%
11.8%	-0.26%	9.8%	
1911/12 – 1920/21	-0.88%	13.0%	-0.97%
13.0%	-0.40%	8.0%	
1921/22 – 1930/31	-0.08%	2.6%	-1.09% ***
2.6%	-0.41%	2.4%	
1931/32 – 1940/41	0.24%	4.0%	-1.16% **
4.0%	0.10%	3.0%	
1941/42 – 1950/51	-0.53%	4.2%	-1.78% ***
4.2%	-1.45% **	4.0%	
1951/52 – 1960/61	4.24% ***	5.1%	2.28% ***
5.1%	2.34% ***	4.2%	
1961/62 – 1970/71	2.53% **	8.8%	0.32%
8.8%	1.89% **	7.2%	
1971/72 – 1980/81	2.62% **	7.1%	0.41%
7.1%	2.12% ***	5.6%	
1981/82 – 1990/91	3.21% ***	6.2%	1.09%
6.2%	3.23% ***	3.7%	
1991/92 – 2000/01	1.68% ***	3.5%	-0.26%
3.5%	1.65% ***	2.3%	
1901/02 – 1946/47	0.48% ***	8.5%	-0.34% ***
9.6%	-0.01%	6.3%	
1947/48 – 2000/01	2.80% ***	6.5%	0.68% ***
6.7%	2.21% ***	6.3%	

Source: Kurosaki (1999) and update estimates by the author.

Note: "Growth rate" shows a parameter estimate for the slope of the log of Q (or Q/L or Q/A) on a time trend, estimated by OLS. The parameter estimate is statistically significant at 1% ***, 5% **, or 10%* (two sided t-test). "Coeff. var." shows the coefficient of variation approximated by the standard error of the OLS regression.

Table. Growth Performance of Pakistani Agriculture

value)	**Q/L (Output per capita)**		**CQ/A (Output per acre)**
Period	**Growth rate**	**Coeff. var.**	**Growth rate**
Coeff. var.	**Growth rate**	**Coeff. var.**	
1901/02 – 1910/11	4.32% **	15.1%	2.75%
15.1%	0.99%	10.9%	
1911/12 – 1920/21	-0.33%	14.6%	-1.18%
14.6%	-0.19%	6.6%	
1921/22 – 1930/31	-0.64%	10.3%	-1.73%
10.3%	-1.15%	7.4%	
1931/32 – 1940/41	2.81% ***	5.8%	0.97%
5.8%	1.86% **	5.1%	
1941/42 – 1950/51	0.05%	6.7%	-2.92% ***
6.7%	-0.19%	3.4%	
1951/52 – 1960/61	3.44% ***	5.2%	1.00%
5.2%	1.66% ***	3.3%	
1961/62 – 1970/71	5.85% ***	5.2%	2.99% ***
5.2%	3.93% ***	4.5%	
1971/72 – 1980/81	3.24% ***	3.7%	0.09%
3.7%	1.75% ***	3.3%	
1981/82 – 1990/91	3.50% ***	5.2%	0.85%
5.2%	2.65% ***	5.3%	
1991/92 – 2000/01	2.30% ***	5.3%	-0.35%
5.3%	1.61% ***	5.5%	
1901/02 – 1946/47	1.30% ***	12.8%	-0.03%
11.9%	0.38% ***	8.6%	
1947/48 – 2000/01	3.58% ***	7.3%	0.77% ***
7.5%	2.34% ***	6.3%	

Crop Mix

An important aspect of agricultural transformation is changes in crop composition and diversification/specialization. Even when there is no change in land productivity of individual crops in each location, aggregate land productivity can grow through spatial shifts of crops, which reflect the development of more efficient rural markets.

The spatial shifts are the result of farmers' attempts to diversify farming activities towards non-traditional, high value-added crops such as fruits and vegetables and towards livestock activities (Chand 1999).

To capture the long-term changes in crop mix, the Herfindahl Index of crop acreage was calculated. Let *Si* be the share of crop *i* in the sum of the principal crops covered in this chapter, in terms of acreage. Then the Herfindahl Index is defined as

(2) *H* .... "*iSi*,

which has an intuitive meaning of a probability of hitting the same crops when two points are randomly chosen from all the land under consideration.

To investigate whether or not these changes in crop mix were consistent with those indicated by comparative advantage and market development, changes in aggregate land productivity were decomposed into crop yield effects, static inter-crop shift effects, and dynamic inter-crop shift effects (Kurosaki 2003). Let *Yt* denote per-acre output in year *t*, *i.e.*, *Q/A*. Its growth rate from period 0 to period *t* can be decomposed as

(3) (*Yt-Y*0)/*Y*0 = ["*iSi*0(*Yit-Yi*0) + "*i*(*Sit-Si*0)*Yi*0
+ "*i*(*Sit-Si*0)(*Yit-Yi*0)]/*Y*0,

where subscript *i* denotes each crop so that *Yit* stands for per-acre output of crop *i* in year *t*. The first term of equation (3) captures the contribution from productivity growth of individual crops. The second term shows 'static' crop shift effects since it becomes more positive when the area under crops whose yields were initially high increases relatively.

The third term shows 'dynamic' crop shift effects because it becomes more positive when the area under dynamic crops (*i.e.*, crops whose yields are improving) increases relative to the area under non-dynamic crops. Decomposition results for India.

Throughout the post-Partition period, the contribution of both static and dynamic crop shift effects were substantial. More interestingly, during the 1990s, the growth due to improvement in crop yields was reduced while the growth due to static crop shifts maintained the level of the 1980s.

As a result, relative contribution of static shift effects was as high as 31% in the 1990s. This is the highest among the post-independence decades. Therefore, we can conclude that the changes in crop mix in the 1990s were indeed consistent with comparative advantages of Indian agriculture so that the aggregate land productivity was improved. In contrast, the crop concentration index in Pakistan did not show acceleration in the 1990s. Rather it remained at the highest level, which was already reached during the 1980s. This seems to indicate that acreage shifts towards crops with comparative advantages occurred earlier in Pakistan than in India, possibly reflecting Pakistan's earlier attempt to liberalize agricultural marketing during the early 1980s. Decomposition results for Pakistan, the contribution of both static and dynamic crop shift effects were substantial and their shares were larger than in India, throughout the post-Partition period.

This indicates that crop shift effects are more important in a smaller economy. During the 1990s, the growth due to improvement in crop yields declined substantially while the growth due to static crop shifts recovered. As a result, relative contribution of static shift effects was 25% in the 1990s, which was higher than the post-independence average (14%). Here is a similarity between India and Pakistan — in both agricultures, the crop shifts were an important source of land productivity growth in the 1990s.

LONG-TERM TRENDS AND AGRICULTURAL POLICIES

Total output growth rates turned from zero or little to significantly positive levels, which were sustained throughout the post-independence period; crop mix changed with increasing concentration since the mid 1950s.

Therefore, as far as the data suggest, agriculture in both countries has been experiencing one-way concentration of crops since the mid 1950s, when agricultural transformation in terms of output per agricultural worker was proceeding. According to Timmer's (1997) stylization, this is a stage before a mature market economy with diversified production and consumption at the national level. Against our initial expectation, these trends were not reversed in the 1990s in the direction of more diversified production if we focus on major crops.

The trends were strengthened in India during the 1990s while they became less significant in Pakistan. The performance in the 1990s suggests that agriculture in India has remained in the phase of diversification where concentration into high value-added crops is strengthened.

To investigate the impact of institutional and policy changes in the 1990s, we compare major changes in political institutions compiled, with long-term trends in production. The colonial period in the early twentieth century was associated with free trade of agricultural commodities. Under this institutional setting, public investment into irrigation contributed to the rapid expansion of production, especially in regions currently belonging to Pakistan. The colonial period ended with increasing state control on trade and marketing under the war pressure. This period was associated with stagnation in agricultural production.

Policy shifts in agricultural development after independence were indeed similar in India and Pakistan if we look back from a long-term perspective. In the early 1950s, the focus was on land reforms on the one hand and on irrigation development on the other hand. Land reforms were, however, not implemented thoroughly, leaving the disparity between the landed and the landless almost unchanged. Since the late 1960s, the policy emphasis moved to technological innovation called the "Green Revolution" in both India and Pakistan. The rapid increase in land productivity of wheat and rice thanks to the Green Revolution is well documented. As emphasized by Kurosaki (1999, 2002), it is of interests to note that a substantial increase in aggregate land productivity occurred *before* the introduction of high-yielding varieties of rice and wheat, mainly through crop shift effects quantified.

These periods were followed by institutional and policy changes in the direction of liberalization, deregulation, and privatization in India and Pakistan, which are continuing today. However, if we look at the individual components of the policy reform, there are several contrasts between India and Pakistan in the 1980s and 1990s. In India, policies to promote rural employment and reduce poverty were in force since the late 1970s. Most of them were not abandoned but revised under the Economic Reforms since 1991. In contrast, major liberalization policies in agriculture were adopted since the early 1980s in Pakistan with little emphasis on rural employment or poverty eradication policies.

The contrast in crop mix dynamics in the 1990s in India and Pakistan is suggestive. The crop concentration index did not show acceleration in the 1990s in Pakistan while it accelerated in India, but the concentration indices of both countries were at the highest level in the 1990s in the whole 20th century. The similarity could be attributable to the general effects of liberalization, deregulation, and privatization policies in agriculture. The contrast could be attributable to the earlier adoption of these policy reforms in Pakistan. Pakistani farmers were exposed to international relative prices earlier than Indian farmers were. In other words, Indian agriculture seems to have more room for productivity growth through crop shifts today than Pakistani agriculture has. In this aspect, the implementation of WTO agreement on agriculture in India since 1995 is expected to reinforce this advantage. However, as shown by Chand (2003), the WTO implementation has worsened the overall terms of trade for agriculture in the late 1990s, thereby blocking the full exploitation of this advantage.

The comparison of our results with those reported by Misra and Rao (2003) is also interesting. They showed that agricultural export was the main source of growth for Indian agriculture in the 1990s, which was facilitated by trade liberalization policies and the devaluation of Indian Rupee. Their finding is reinforced by our results since we have shown that crop shifts are an important source of productivity growth in the 1990s and shifts to export crops are consistent with out results. Although our analysis has shown the importance of crop shifts in improving aggregate land productivity, it is an underestimate of their true impact because our measure only covers eighteen major crops. Under the Economic Reforms, contract farming is gaining its importance in India in the 1990s (Singh 2002). Similar changes are observed in Pakistan as well, although not so strong as in India. Our measures of the impact of liberalization policies on output growth in the 1990s should be understood as an underestimate because the growth in non-traditional crops such as vegetables or horticultural crops was not captured. Based on a production dataset that corresponds to the current border of India and Pakistan for the period c.1900-2000, we investigated the performance of agriculture in the two regions and associated it with changes in political institutions and agricultural policies. Through examining the growth records of agricultural production and changes in crop mix indices, we showed that institutional and policy changes have significant effects on agricultural growth in India and Pakistan. Farmers have responded to these changes, adjusting their crop mix and production technology. The crop concentration indices were at the highest level in the 1990s both in India and in Pakistan, suggesting that liberalization/deregulation/privatization policies in agriculture in these countries led to an increased role of crop shifts in enhancing aggregate land productivity.

To quantify structural determinants of these changes and net effects on the welfare of rural population, further research is needed, such as analysis of production costs, investigation of minor crops and livestock activities, etc. These are left for future study. Instead, implications for agricultural policies in the 21st century are explored to conclude this chapter. First, institutional and policy changes are likely to have significant effects on agricultural production in India and Pakistan. This is confirmed by the fact that the sustained growth in the total output began just after the Partition, well before the introduction of the Green Revolution technology, and also by changes of these trends in the 1990s. The current controversies in India and Pakistan on contract farming, corporate farming, and land ceiling legislation should be viewed from this historical perspective.

Second, farmers in India and Pakistan have responded to the changes in market conditions so that they not only adopted new technology with high-yielding potentials but also adjusted their land allocation towards high value crops. The importance of the effects of land re-allocation should not be ignored. For example, the effects of reforms in the price support system of major crops cannot be isolated from their interactions with crops whose prices are not under the support price system.

Third, although not discussed fully in this chapter, existing evidence suggests that public investment in/for agriculture were reduced in the 1980s and 1990s. This occurred both in India and Pakistan under the banner of "Economic Reforms" or "Structural Adjustment." It should be emphasized that the sustained growth during the post-colonial period was achieved when substantial public investment was implemented. With reduced public investment without simultaneous improvement in investment efficiency, the boom experienced during the 1990s in response to newly opened opportunities would not last long. The importance of production-oriented infrastructure in increasing productivity of agriculture and in reducing rural poverty cannot be overemphasized. Rather, considering the public-good nature of such investment, its importance should rise under the context of globalization and trade liberalization.

CHALLENGES OF INDIAN AGRICULTURE

Indian agriculture is facing a policy paradox. Although several forecasts of the 1990s predicted that India would be a large importer of grains in the years to follow, in fact from 2001 to 2004 India exported around 30 million tons of foodgrains. It was seeking primarily to liquidate its bulging grain stocks, which reached 63 million tons in July 2002. Whereas India's agricultural policy is still rooted in the goal of self-sufficiency in grains, consumption patterns are changing fast towards high-value agricultural products such as fruits and vegetables, livestock products, and fish.

The policy environment is lagging behind the structural change occurring in India's consumption and production baskets. On another front, foreign exchange reserves, which had reached a rock-bottom US $1.2 billion in July 1991, climbed to more than US $120 billion by the end of 2004. Nonetheless, despite comfortable food and foreign exchange reserves and reasonably high growth in Gross Domestic Product (GDP) of about 6 per cent annually, India still has more than 250 million underfed people and high under-employment.

This situation reflects severe problems on the distribution front. What are the reasons behind this paradoxical situation? The answer presumably lies in the neglect of, as well as misallocation of resources in, agriculture and rural development, especially in the later phase of the reform process initiated in 1991. The average annual rate of growth in agriculture fell from more than 4 per cent per year during 1992/93 to 1996/97 to less than 2 per cent per year during the period 1997/98 to 2002/03, and it remains low. What led to this dramatic decline in the growth of agriculture since 1997/98?

How can it be revived? How can growth in agriculture and rural development diminish poverty quickly? To stimulate pro-poor agricultural growth and rural development, India will need to make some strategic choices. We propose action in five major areas that can help the government to accelerate agricultural growth and reduce poverty, malnutrition, and unemployment quickly and on a sustainable basis. All of these reforms can be achieved with due regard for the well-being of the country's rural poor.

AGRICULTURAL CHANGE IN PLANTINGS

Agricultural change refers to the difference between the first plantings 10,000 years ago and today's computerised, industrialised, genetically engineered production systems; but a process that occurs on a daily basis, as farmers make decisions about what, where, and how to cultivate. It goes well beyond how much food is produced, how much money is made, and how the environment is affected: Agriculture is intimately linked to many institutions in every society, and to population.

Agricultural change is mainly based on sustainable livelihoods and agricultural intensification. A livelihood is sustainable if it can cope with, and recover from, stresses and shocks, maintain or enhance its capabilities and assets and provide net benefits to other livelihoods locally and more widely, both now and in the future, without undermining the natural resource base. Agricultural intensification is increased average inputs of labour or capital on a small–holding, either cultivated land alone, or on cultivated and grasing land, for the purpose of increasing the value of output per hectare.

It may occur as a result of an increase in the gross output in fixed proportions due to inputs expanding proportionately, without technological changes, a shift towards more valuable outputs or technical progress that raises land productivity. However a shift from a high yielding crop of 100 days per year to two low-yielding crops of 250 days per year cannot be described as intensification, though it may shift output value towards labour. The international agricultural development policy based on CGIAR denies developing world small-holders and nations an opportunity to diversify food production, adapt new improved local techniques of production and resist global market competition.

Through this international concept of development, techniques and resources of the developing world are devalued. An increase in living standards is hinged on the production of cheap food and non-food agricultural products, for global market competition at the expense of social and ecologic characteristics. Globalisation has come at the price of undermining some of these characteristics like diversity, adaptability, and resilience in exchange for the overreaching goal of efficiency.

Meanwhile, citizens of these externally dominated nations starve as tons of flowers are delivered to other states from their soil. Unless one adopts a coldly Malthusian perspective, the situation is worthy of examination and further study. Aside from the degradation of human life imposed by economic agricultural development policy, there are informational costs. The loss of diversity, biotic and cultural, further degrades the quality and quantity of information available regarding natural farming techniques. As indigenous farming knowledge gives way to chemical sprays and imported seeds, not only the perceived, but actual value of that knowledge is diminished. Onfarm recycling is necessary for locally sustainable farming and contradicts the conventional high-external input approach.

Local food markets provide a means to confine inputs and outputs in the current context of needs and sustainability of the community. In order for these outside forces to work in balance with the farm, there must be shared goals.

The conflict between the state and the individual arising within the modern agricultural paradigm is the result of conflicting agendas between the top and the bottom.

Change Process Affecting Agricultural Production and Livelihoods

The lack of transparent, timely and reliable crop and livestock marketing information is seen by many as one of the greatest challenges to the development of the agriculture industry in East India. Without access to information about agricultural products prices at different regional markets, farmers are unable to identify which points-of-sale offer the best prices for their agri-products. For example, in Mbirikani, Kajiado County, Kenya, Maasai pastoralists once relied on middlemen to gain market information.

However, a marketing information system has been developed by the Livestock Information Network and Knowledge System (LINKS) to support livestock producers and traders. Now the farmers use mobile phones to ask for livestock prices in Emali, Mombasa and Nairobi, and this knowledge significantly improves their bargaining power with livestock traders.

Increasing Marketing Opportunities

Given the high dependency of pastoralists' livelihoods on the sale of livestock and livestock products, LINKS designed and implemented an Information Communication Technology (ICT) infrastructure to collate data on livestock sales and prices from a network of different markets for dissemination using SMS messages.

This has been a success and has been used to develop a National Livestock Marketing Information System (NLMIS) in Kenya. NLMIS relies on a network of County livestock marketing officers and data monitors to transmit data between markets and the database. The field officers are also trained to download, analyse, and summarise the informa-tion to transmit to pastoralists and traders, including info-rmation about prices and volumes of cattle, camels, sheep and goats at livestock markets.

The market information is also made available online, and may be downloaded for printing and sharing with livestock communities with no internet access. The information is also shared by e-mail, posted on billboards at market places and can be requested through SMS.

Vibrant Markets

This change in communication has also transformed undeveloped markets in other value chains in agriculture. The availability of marketing information through SMS and information boards, competition among agri-products traders has increased, and this has improved and stabilised prices paid to farmers.

Women have also benefited by providing animal health services, fodder and small loans to the traders.

CHALLENGES IN PRODUCTION FOR RURAL LIVELIHOODS AND OPPORTUNITIES

Some of the challenges that affect production in the rural livelihoods in developing states, especially in India include:

Climate Change

Climate change is exacerbated by other factors like widespread poverty, recurrent droughts and floods. Others are dependence on natural resources and biodiversity, over–dependence on rain-fed agriculture, heavy disease burden, and numerous conflicts that have engulfed the continent. Changes in future climate may affect negatively the overall economy of India, hence hampering its potential economic growth.

Bio-fuel

From living organisms or the waste these are produced, especially from crops such as maize, soya, sugarcane and rapeseed.

These bio-fuel are of two types:

1. Bioethanol made from sugarcane, sugar beet and cereal crops, and
2. Bio-diesel made from soybean, rapeseed, vegetable oils, animal fats used as frying oils.

The question being asked is:

- Is bio-fuel production a solution to climate change or cause of global hunger? If made from plants, is it environmental friendly as opposed to fossil fuels. It is argued that crops absorb carbon dioxide from the atmosphere as they grow and produce about 60 per cent less carbon dioxide than fossils. But land previously used for food crops is being turned over to bio-fuels.

Population and Migration

Population growth is an important factor for local environment change. Rapid population growth is one of the main reasons for increasing the number of people. The move for livelihood and migration has been a major factor of rapid population growth in urban areas in less developed states.

Urbanisation

It is the driving force for modernisation, economic growth and development. However, there is increasing concern about the effects of expanding cities on human health, livelihoods and the environment. The implications of rapid urbanisation and demographic trends for employment, food security, water supply, shelter and sanitation, especially the disposal of wastes that the cities produce are staggering. The question that arises is whether the current trend in urban growth is sustainable considering the accompanying urban challenges such as unemployment, slum development, poverty and environmental degradation, especially in the developing states.

Role of Agricultural Information

Although agriculture is the backbone of the economies of most Indian states, development in this sector has not grown as fast as the population. Women in India are the key actors in agriculture; and produce about 80 per cent of the region's food. The major problems include cultural, social, economic, legal, education and lack of information to improve farming activities. Lack of reliable and comprehensive information is one of the major hindrances to agricultural development.

Population growth posses the following challenges in production and livelihood opportunities: Intensification of agricultural production, soil degradation, *e.g.,* erosion, lack of access to land, reduced livestock grasing possibilities, and limits to grasing land. Others challenges include degradation of water resources, difficult access to irrigation water, decreased size of food portions and declining nutrition, deteriorating health standards, problems accessing credit, and difficulty in accessing education and training. Still others include demographic trends towards a younger population places increasing pressure on labour markets, resulting in greater unemployment and out-migration, increased political instability and conflicts, increases affluence gap between rich and poor regions which has implications for migration, and it fuels the desire to emigrate from poor regions.

Yet there is increased potential for demographically fuelled conflict resulting from declining livelihood assets. Further, environmental problems include destruction of natural ecosystems, increased rate of species extinction, falling water tables and depletion of aquifers, pollution of rivers, seas and coastal waters, and increase of harmful emissions to the atmosphere.

CHALLENGE TO PRODUCTION AND RURAL LIVELIHOOD OPPORTUNITIES

Men and women suffer similar problems, but these seem to affect women more adversely. Unfortunately, access to information has not received adequate attention in most states and especially in rural aresas where 70–80 per cent of the Indian population live. Women in rural areas have very little access to information, and most are poor and cannot read or write. Rural and peri-urban women, therefore, need to be empowered to increase agricultural productivity through access to information. The few assessments that have been conducted on the information needs of women in agriculture have shown that women require information on all aspects of agricultural production, processing, marketing, decision-making processes, the resource base, and trade laws. They also need to exchange information on indigenous knowledge and require appropriate Information Communication Technologies (ICTs) to be able to access vital information efficiently and cost-effectively.

The Role of ICT as a Communication Tool

Traditional ICTs such as drama, dance, folklore, group discussions, meetings, exhibitions, demonstrations, visits, farmer field schools, agricultural shows, radio, television, video and print media have been used successfully in many Indian states. The media play a major role in delivering agricultural messages, with the radio forming the main source of information. Satellite, solar and fibre optic technologies are now in use for computers, telephones and facsimile. Where appropriate, these should be tapped to enable rural women farmers to access information using modern ICTs concurrently with traditional ones.

Telecentres are the way forward and could be established in villages and new ICTs such as electronic mail, World Wide Web, electronic networks, Newsgroups, ListServs, teleconferencing, CD-ROM and distance learning tools can be used where appropriate by rural farmers, farmer leaders or their intermediaries such as non-governmental organisations, community based organisations and development agencies. These intermediaries would in turn repackage the information for the consumption of rural farmers including women.

The agricultural information sector has a relatively weak technological base and insufficient scientific, technical and educational capacity and thus needs substantial capacity building.

There is also a dire need for radical change in agricultural information systems, if widespread management and dissemination of information and knowledge in digital format is to take root. This change can be driven by building partnerships among agricultural information professionals in the region, the government and international and regional partners such as FAO, AARINENA/ RAIS, ICARDA, AOAD and ASARECA.

Productive Capacity and Change of Agro-ecosystems as a Challenge to Livelihood Opportunities

Agro–ecology is a scientific discipline that uses ecological theory to study, design, manage and evaluate agricultural systems that are productive but also resource conserving. It considers interactions of all important biophysical, technical and socio–economic components of farming systems. Broadly stated, it is the study of the role of agriculture in the world. Agro–ecology also provides an interdisciplinary framework with which to study the activity of agriculture.

In this framework, agriculture does not exist as an isolated entity, but as part of ecology of contexts. Agro–ecology draws upon basic ecological principles for its conceptual framework. It refers to an ecological approach to agriculture that views agricultural areas as ecosystems and is concerned with the ecological impact of agricultural practices. Agro–ecosystems on the other hand, refer to a model for the functioning of an agricultural system, with all inputs and outputs. An ecosystem may be as small as a set of microbial interactions that take place on the surface of roots, or as large as the globe.

An agro ecosystem may be at the level of the individual plant-soil-microorganism system, at the level of crops or herds of domesticated animals, at the level of farms or agricultural landscapes, or at the level of entire agricultural economies.

They analyse; mineral cycles, energy transfo-rmations, biological processes and socioeconomic relatio-nships as a whole in an interdisciplinary fashion. Further, they are concerned with the maintenance of a productive agriculture that sustains yields and optimises the use of local resources while minimizing the negative environment and socio-economic impacts of modern technologies.

In industrialised states, modern agriculture with its highinput technologies generates environmental and health problems that often do not serve the needs of producers and consumers. In developing states, addition to promoting environmental degradation, modern agricultural technologies have bypassed the circumstances and socio-economic needs of large numbers of resource-poor farmers.

Agro-ecology is the use of ecological theory to evaluate agricultural systems that are productive and resource conserving. It encompasses interactions of biophysical, technological and socio-economic components of farming systems. It is characterised by energy and nutrient flow, biodiversity and evolution.

The main challenge to agroeco-systems is the modern technology and may also include social, cultural and economic aspects of life. The main drivers of the agro-ecosystems change are agricultural expansion, intensification of production and environmental concerns including climate change.

Policy issues are of paramount importance. Therefore, Ecological agriculture aims at improving agricultural production and post-production while conserving the regenerative and reproductive capacity of the natural resource base, and thus, avoiding the cycle of rectification of error.

It is based on traditional ecological understanding and combines results of modern science of natural processes. Hence, ecological agriculture is about maximizing use of knowledge of natural processes. Criteria for ecological agriculture need to consider technical performance, efficiency, impact, resource use, availability, and user preference.

Climate Change and Shift to Bio-fuels: A Challenge to Production and Livelihoods

Livelihood comprises of the capabilities, and material and social assets necessary for a means of living. On the other hand, Climate change is the natural cycle through which the earth and its atmosphere are going to accommodate the change in the amount of energy received from the sun. The climate goes through warm and cold periods, taking hundreds of years to complete one cycle. Changes in temperature also influence the rainfall, but the biosphere is able to adapt to a changing climate if these changes take place over centuries.

Climate change poses a serious threat to agriculture, particularly in developing states in Asia and India. Both climate variability and extreme climatic events are affecting agricultural productivity and food security in Asian and Indian states.

Moreover, climate change impacts are affecting the livelihoods of common people and aggravating global poverty. Millions of poor across the Asia and India derive their livelihoods from common property and natural resource bases like land, water, fisheries, and forestry. Are bio-fuels a solution to climate change or cause of global hunger? Researchers have concluded that shift to bio-fuels is a mistake. It will lead to clearing forests to plant crops for fuel and this will release gases. It is said that increasing production of bio–fuels to combat climate change will release between 2-9 times more carbon gases over the next 30 years than fossil fuels developed states should re-direct bio-fuel subsidies towards preventing deforestation for a better climate outcome. Clearing forests produces an immediate release of carbon gases into the atmosphere followed by loss of habitats, wildlife and livelihoods.

The demand on arable land cannot be met in the EU or the US to plant bio-fuel crops. There is high possibility to shift the burden on land in developing states, *e.g.*, Brazil, Paraguay, Indonesia, which have huge deforestation programmes to supply the world bio-fuel market. Any delay in addressing climate change through reduction of greenhouse gas emissions will increase the cost of both adaptation and protecting development. Severe food insecurity and poverty in many states is likely to deteriorate significantly as a result of climate change. Climate change, agriculture and food security, natural resources and environmental sustainability, economic growth and poverty reduction are interdependent and this interdependence needs to be at the core of comprehensive policy planning and an integrated resource allocation and implementation strategy, both spatially and temporally.

DEVELOPMENT PARADIGMS AFFECTING AGRICULTURE AND LIVELIHOODS

Small-holder agriculture in India has consistently under-performed, for reasons that remain only partly understood, despite a succession of theoretical paradigms and analytical frameworks that have been translated into policy prescriptions which have similarly failed to deliver sustained and significant increases in agricultural yields. For example, the radical reforms implemented under agricultural liberalisation programmes throughout India in the 1980s and 1990s were grounded in a theory that agriculture was being stifled by excessive interventionism by the state in agricultural production and marketing, while farmers and traders were undermined by unsustainable input and output subsidies.

The shift from "state-led" to "market-led" agriculture achieved disappointing results, leading to a reassessment of both the diagnosis and policy prescriptions. There were several explanations why agricultural liberalisation failed in India. These ranged from deficits in the necessary enabling environment, infrastructure like information systems and markets. The states had also withdrawn from research and extension services. However, in other developing states like in Asia, the states withdrew from agriculture after the necessary investments had been made in infrastructure and market development.

An alternative explanation was that liberalisation failed because it was imperfectly and incompletely implemented. Another view was that small-holder agriculture in India has under-performed because of lack of protection or recognition of small-holder farmers by the policy makers and the international bodies like the UN during their "modernisation" phase.

This protection was necessary to avoid unfair competition from subsidised products from wealthy industrialised states. After a number of debates, certain policies have been put in place to revitalise the agriculture sector in India and especially SSA.

They include:

- Improve access to markets and develop modern market chains;
- Assist subsistence farmers to enter markets, and foster sustainable resource management;
- Achieve food security and improved livelihoods for those who remain as subsistence farmers, including improving the resilience of farming systems to climate change; and
- Capitalize on agricultural growth to develop the rural non-farm sector.

Some steps to achieve these outcomes have been suggested as including improvement in agriculture market chains, expanding exports, increasing uptake of yield-stabilising technologies, and reducing weather risk.

Agricultural Technology

- To satisfy the increasing needs of food security in India, especially SSA, substantial technological innovation should take place. To increase animal and crop production, modern biotechnology tools of recombinant DNA, including genetic engineering, offer some opportunities for generating such innovation.

- Though the use of some biological technologies has met resistance on social and ethical requirements, biotechnology offers the best solution to reduce hunger in India. Genetically engineered products like GM crops has met with fierce resistance, particularly in Europe on ethical grounds and on concerns of perceived negative impacts of GM crops on the environment and food safety. Drugs and vaccines have not stirred much controversy.
- Ethical considerations revolve around topics such as the "unnatural" nature of gene transfers across species; possible socio-economic impacts of widening the gap between the rich and poor farmers and states and the fear that agricultural biotechnology will increase the dependency of global food supply on few multinational corporations controlling the seed industry.
- With all the negative publicity of GM crops, there are concerns that the resistance to GM crops by consumers in Europe may have hindered the transfer of this new innovation to the developing states where increasing crop productivity is most urgent.

Despite this resistance, biotechnology is already in play and necessities of life like food, feed, fibre, fuel and medical drugs are being produced. In agriculture, biotechnology tools have been used for animal and plant disease diagnostics, for production of recombinant vaccines against animal diseases and for the improvement of livestock and crops. There has been some lack of support for the development of GMOs in the fight for food security in India from some sectors.

However, National Academies of Sciences called for a concerted effort by all sectors, public and private, to develop GM crops, especially food staples that will benefit consumers and poor farmers in developing nations. It called for the sharing of GM technology developed by private corporations for use in hunger alleviation and to enhance food security in developing states. The role of biotechnology involves the use of living organisms for human benefit.

It consists of two components:

1. Tissue and cell culture and
2. DNA technologies including genetic engineering.

Both components are essential for the production of GM plants and animals. Plant tissue and cell culture are relatively low cost technologies which are simple to learn, easy to apply and widely practiced in many developing states.

Plant tissue culture aids crop improvement through a range of actions, including:

- Mass propagation of elite stock;
- The provision of virus-free stock through *in vitro* culture of meristem;

- The selection and generation of somaclonal variants with desirable traits;
- The overcoming of reproductive barriers and the transfer of desirable traits from wild relatives to crops by wide crosses;
- The facilitation of gene transfers using plant protoplast fusions;
- Anther culture to obtain homozygous lines in a breeding programme; and
- *In vitro* conservation of plant germplasm.

The second component of biotechnology, *i.e.,* DNA technologies, including genetic engineering, utilises newly emerging knowledge of the genes and the genetic code to improve crops, trees, livestock and fish. In order to fully realise the benefits of agricultural biotechnology to contribute to poverty alleviation and equitable growth in developing nations, it is crucial that a concerted effort be made to ensure that the benefits of biotechnology are available to a broad spectrum of small farmers in a range of developing states.

For example, the sharing of the rice genome sequence data with researchers in developing states by private sector and the recent announcement of the sequencing of the banana genome are steps in the right direction towards using biotechnology for the benefit of developing nations. Also, there is the question of assisting in capacity building in biosafety in compliance with the recently agreed Cartagena Protocol on biosafety regulations. This is because the owners of proprietary technologies would be very reluctant to licence their technologies in states that have no biosafety legislation and controls in place.

The development of biosafety in developing states is also important in cases where these states may have to import GMOs for use in breeding programme/ testing in their environment and/or for use as food and feed. Facilitate access to proprietary technologies to developing states. The issue of Intellectual Property Rights (IPR) needs to be addressed. It is important that capacity building in IPR is stimulated in order to meet the minimum requirements of the WTO-TRIPS.

ROLE OF INFORMATION TECHNOLOGY IN AGRICULTURE

The agriculture scenario all over the world is undergoing a rapid change particularly after WTO Agreements came to existence. In order to take full advantage of the changing global agricultural scenario interconnecting of policies related to pricing, marketing and trading of agricultural commodities are also reviewed. Simultaneously there is also need to review and revitalise the mechanism for transfer of technology under changing environment. It is readily accepted that increased information flow has a positive effect on the agricultural sector and individual firms. However, collecting and disseminating information is often difficult and costly. Information Technology (IT) offers the ability to increase the amount of information provided to all participants in the agricultural sector and to decrease the cost of disseminating the information. An understanding of the factors associated with IT adoption and use in agriculture will enable the development of strategies to promote IT adoption and increase the effectiveness and efficiency of information used in agriculture.

It is a fact that access to information holds the key for successful development. Improved communications and information access is directly related to socio-economic development of any nation. Agriculture is one of the prospective areas in which IT can effectively be applied particularly for the social and economic development of the Indian agrarian community. However, rural population in our country still have difficulties in accessing crucial information in the forms they can understand in order to make timely decisions for better farming. IT is generating possibilities to solve such problems of different categories of end users. For this purpose electronic communications infrastructure needs to be established in the country for remote rural areas. The challenge is not only to improve the accessibility of communications technology to the rural population but also to improve the relevance of information to local development. The present object depicts the changing scenario of information dissemination exploiting the information technology to the farmers for their agricultural development.

IT for Agricultural Production and Marketing

IT is playing an important and vital role in agricultural production and marketing. IT allows farmers to save time on order and delivery and getting feedback. In the existing competition, there is a need to rapidly attract new customers as well as retain existing customers. In order to take the real status of agricultural production and marketing.

There is an urgent need to develop the following items:

- *Farmers' Crop Database must be Managed*: The database includes the kinds of crops, the size of cultivated area, time of harvest and yield. Farmers or the extension personnel transmit those data via the Internet to database server. Further, information provides the farmer with an important instrument for decision making and taking action.
- *Crops Information Service System should be Created:* This system analyses the crop data to create some statistical tables. Farmers can access these statistical data by browsing the homepage and make their production plan. Changes within the structure of agriculture will probably have an impact on the selection and types of acquisition of software and other integrated systems made by the farmers.
- *Production Techniques and Information Enquiry System should be Created*: This system integrates the production techniques and information, which are developed by experimental agricultural institutes and agricultural improvement stations. Farmers can find

out relevant production information through this enquiry service system.

- *Production Equipment's Enquiry Service System should be Created*: This system gathers information from the companies of seeds and crop production equipment to build the production equipment's enquiry service system. At the same time, allow relevant companies to access this system and enter their own data. Therefore, farmers can order the needed items through this system.

Information is critical to the social and economic activities that comprise the development process. Development economy has witnessed for revolutions in agricultural, bio-technological, industrial and information technology. Good communication system and information system reinforce commitments to sustainable productivity.

The Government of India is giving more thrust on agriculture, food and information technology sectors towards achievement of economic reforms to achieve high growth rate in production in the years to come. The National Agriculture Policy announced addresses the challenges arising out of economic liberalisation and globalisation. It seeks to actualise the vast untapped growth potential of Indian agriculture, strengthen rural infrastructure to support faster agricultural development, promote value addition and secure a fair standard of living for farmers and farm workers.

The National Agriculture Policy lays emphasis on the use of Information Technology for achieving a more rapid development of agriculture in India. In pursuance thereof, the Department of Agriculture and Cooperation (DAC) has formulated information technology (IT) Vision 2020.

This vision inter alia envisages that:

- Information relating to agriculture sector would be available to the ultimate users—the farmers—for optimising their productivity and income;
- Extension and advisory services making use of information technology would be available to the farmers on round the clock basis;
- The tools for information technology will provide networking of agriculture sector not only in the country but also globally and the Union and State Government Departments will have reservoirs of data base; and
- The long term vision on "Information Technology in Agriculture Sector" is to bring farmers, researchers, scientists and administrators together by establishing "Agriculture on-line" through exchange of ideas/information.

In future, information technology will reduce the cost and time of information system. IT will bring new information services for agricultural development that will enable the farmers to have much greater control over the information channels.

Agricultural Implications of IT

Information technology provides answers to a number of questions to the farmers. For example, what are the benefits of more irrigation? Is it cost-effective to apply additional chemicals? When is the best time to sell crops or buy inputs? With improved record-keeping, more detailed cost analysis and more sophisticated marketing strategies, farmers are making better decisions and earning higher profits. The Internet is increasing communication and business opportunities within the agricultural community, which previously operated in the relative isolation of rural areas.

Farmers, agricultural researchers, cooperatives, suppliers and buyers use the Internet to exchange ideas and information, as well as to conduct business with each other. Machinery, seed chemicals and other types of agricultural products can be purchased and sold online. People can search for jobs and employees. It is to be noted that, the farmer is in no position to use IT directly. The literacy levels, language barrier as most of the application software are predominantly in English, cost of computers, poor communication infrastructure make it impossible for individual farmers, particularly small farmers to directly adopt IT.

This calls for institutional effort to harness to create the necessary IT based services to farmers. But in India, one prominent problem is that most of the farmers own small holdings, this seems to be difficult. In this situation, it may be made possible by adopting the corporate farming system, which is the need of the hour with advent of new agricultural policy. By taking up corporate farming, a group of farmers can install a computer and any educated young man from that group can undergo training of how to browse the Internet. He can provide the farmers current commodity, analysis reports on world markets and trade for different commodities.

Food market overviews provide valuable information about some of the most important export market. IT can help to provide the information on the likely price distribution of key commodities over the coming years. Such information helps farmers and traders make decisions on when and in what ways to market their grain. Whether, to sell at harvest or store on-farm in anticipation of higher later in the season.

When combined with enterprise budget data, the information can also be used in deciding which crops to produce in the coming season. In order to encourage farmers to obtain best possible price, information on various agricultural output markets is also being provided. The objective of this activity is to provide status of price at different markets to facilitate farmer to move his produce to the market where he can expect better price.

The entire exercise will not be useful unless necessary arrangements are made to ensure that the farmers utilise this facility. The contribution of information technology in bringing down costs, increasing efficiency and improving productivity and thereby contributing to the bottom line needs no special emphasis.

In the fertilizer marketing context, IT can play a major role in efficient sales operations, checking the marketing costs, safeguarding market share and providing efficient customer services. A well conceived IT set up can endow decision makers at all levels with better reflexes to effectively respond to market conditions. IT helps producers monitor and respond to weather variability on a day-to-day basis. Solar-powered weather stations in the field can be hooked up to a farmer's computer to relay information about current air and soil temperature, precipitation, relative humidity, leaf wetness, soil moisture, day length, wind speed and solar radiation. Producers use the Internet to monitor prices quickly and as often as they like.

Farmers from around the world can exchange ideas, post questions and get answers about specific topics. Thus, it is said that the importance of information technology in the field of agriculture is emerging. The challenges of cost intensive, highly technical agricultural technologies are knocking the door to the farmers which is ignited by the globalisation. There is significant shift from agriculture supply driven to demand driven paradigm in new emerging and changing economics policy. It is viewed that future agricultural growth would be information driven. New information must reach to the ultimate user at the fastest speed to harness its potential benefits. Information like seed, water, nutrients and plants protection biological is one of the key inputs for successful farming. Knowledge intensive and precision farming techniques will be the guiding lines for sustainable agriculture in the future.

5

Agriculture in the Changing Global Scenario

Steady globalization of trade has profound implications for future agricultural development. The diversity of India's agro-ecological setting, high bio-diversity and relatively low cost of labour provide potential for agricultural competitiveness in a globalized economy. It is expected that with increasing globalization of markets over the years there will be demands for agricultural intensification. This will also be favoured because of greater backward and forward linkages between agriculture and food industry. Therefore, increase in production and productivity are bound to be strategically important to economy. Intensification will not only favour alleviation of rural poverty but will also improve resource conservation particularly in the small farming sector where farmers can be encouraged to take up organized production of high value crops such as fruits, specialty vegetables, flowers medicinal and aromatic herbs etc. Stronger demands for crops of the small farmers' will not only improve incomes and welfare but will also make investments in technology and resource conservation more attractive.

The General Agreement on Tariff and Trade (GATT) and liberalization of global trade is bound to have impact on future land use and production pattern. Understanding the local, national and international environment under which agricultural production is taking shape will be crucial in developing our own strategies.

EXTENSION STRATEGIES

Since early fifties a number of public by funded agricultural development programmes have been sponsored. These have included programmes like the National Extension Service (NES) Blocks in 1953, the Intensive Agricultural District Programme (IADP) in 1961-62, the Intensive Agricultural Area Programme (IAAP) 1964-65, the High Yielding Variety (HYV) programme 1966-67 and the Small and Marginal Farmers' Development Programmes (SMFDP) in 1969-70. Though these programmes had a perceptible impact the efforts did not get replicated over different areas and categories of farmers. In mid seventies based on pilot level project in Rajasthan Canal and Chambal command area a 'Training and Visit' (T&V) system of extension was promoted in different states. Extension efforts of the Indian Council of Agricultural Research through its research Institutes and the State Agricultural University were largely limited to demonstration of new technologies through such programmes as National Demonstration Project, Operational Research Project, the Lab to Land Programme and the Krishi Vigyan Kendras. However, there appears much to be desired in the way that extension programmes are conceived and implemented.

At present extension programmes are implemented in largely a top-down fashion leaving little scope for localized planning and action. Farmers are almost passive receivers and their involvement in the process of technology generation and adoption is almost absent. Extension services, at present, are almost exclusively in the public sector domain and there is no effort or institutional support for other operators *e.g.* the NGOs, the corporate bodies etc.

Extension programmes sponsored by the government operate largely in isolation and there appears a strong need to view the extension programmes as an integral part of the research and development process.

The challenges facing agricultural development call for fundamental changes in our approach to technology transfer/extension programmes. Changes are necessary in the context of changing economic environment following policy adjustments in relation to privatization, deregulation and globalization calling for greater efficiency and effectiveness of the extension system. More importantly there is need for

- Greater emphasis on providing producers with knowledge and understanding needed to overcome the problems or to exploit opportunities of their own specific production systems. Correspondingly there will be a need to de-emphasize 'package of practices' or the blanket recommendations, top down approach followed thus far.
- Shift in the focus of public extension systems from promoting inputs use to one on sustainable management of resources and improvements in the production system as a whole.
- Closer interaction between farmers, extension scientists and production system researchers in diagnosing problems and identifying location specific recommendations emphasizing participation and education rather than being prescriptive.
- Widening the range of extension delivering agencies. While the publicly operated extension systems will continue to be important, there will appear a greater role for NGOs, farmers' associations and corporate sectors in particular situations. Role of commercial suppliers of seeds, agrochemicals, machinery, vaccines and medicines in providing advisories, as is already being done in a limited way, will

need to be encouraged and factored into public system's own priorities.

- Wider and more creative use of mass media in tune with current developments in information technology to get information across to the farming community whose ability to overcome constrains at farm level will increasingly depend on access to reliable and up-to-date information.

ACCENT ON EMPOWERING THE SMALL FARMERS

Contributions of small holders in securing food for growing population have increased considerably even though they are most insecure and vulnerable group in the society. The off-farm and non-farm employment opportunities can play an important role. Against expectation under the liberalized scenario, the non-agricultural employment in rural areas has not improved. Greater emphasis needs to be placed on non-farm employment and appropriate budgetary allocations and rural credit through banking systems should be in place to promote appropriate rural enterprises. Specific human resource and skill development programmes to train them will make them better decision-makers and highly productive. Human resource development for increasing productivity of these small holders should get high priority. Thus, knowledge and skill development of rural people both in agriculture and non-agriculture sectors is essential for achieving economic and social goals. A careful balance will therefore need to be maintained between the agricultural and non-agricultural employment and farm and non-farm economy, as the two sectors are closely inter-connected.

Raising agricultural productivity requires continuing investments in human resource development, agricultural research and development, improved information and extension, market, roads and related infrastructure development and efficient small-scale, farmer-controlled irrigation technologies, and custom hiring services. Such investments would give small farmers the options and flexibility to adjust and respond to market conditions.

For poor farm-households whose major endowment is its labour force, economic growth with equity will give increased entitlement by offering favourable markets for its products and more employment opportunities. Economic growth if not managed suitably, can lead to growing inequalities. Agrarian reforms to alleviate unequal access to land, compounded by unequal access to water, credit, knowledge and markets, have not only rectified income distribution but also resulted in sharp increases in productivity and hence need to be adopted widely. Further, targeted measures that not only address the immediate food and health care requirements of disadvantaged groups, but also provide them with developmental means, like access to inputs, infrastructure, services and most important, education should be taken.

Identification of need-based productive programmes is very critical, which can be explored through characterisation of production environment. We have to develop demand-driven and location-specific programmes to meet the requirements of different regions to meet the nutritional security of most vulnerable population in the rural areas. Improved agricultural technology, irrigation, livestock sector and literacy will be most important instruments for improving the nutritional security of the farm-households. Watershed development and water saving techniques will have far reaching implications in increasing agricultural production and raising calorie intake in the rainfed areas. Livestock sector should receive high priority with multiple objectives of diversifying agriculture, raising income and meeting the nutritional security of the poor farm households.

Need based and location-specific community programmes, which promise to raise nutritional security, should be identified and effectively implemented. Expansion of micro credit programmes for income-generation activities, innovative approaches to promote family planning and providing primary health services to people and livestock and education should enhance labour productivity and adoption of new technologies. Development of the post-harvest sector, co-operatives, roads, education, and research and development should be an investment priority. A congenial policy environment is needed to enable smaller holders to take the advantage of available techniques of production, which can generate more incomes and employment in villages.

For this poor farmer needs the support of necessary services in the form of backward and forward linkages. Small-mechanised tools, which minimise drudgery and do not reduce employment, but only add value to the working hours are needed to enhance labour productivity. Special safety nets should be designed and implemented for them. Can agricultural co-operatives internalise and galvanize these marginal and excluded people? Off-farm employment provided through co-operatives will go a long way in pulling them out of the state where poverty breeds poverty.

Therefore, investment in the empowerment of the small landholders will pay off handsomely. Let us create rural centres of production and processing by masses through co-operatives or empowerment of Gram Panchayats to promote co-operatives. This will improve efficiency of input and output marketing and give higher income. There is need to disseminate widely post-harvest handling and agro-processing and value addition technologies not only to reduce the heavy post-harvest losses but also improve quality through proper storage, packaging, handling and transport. Panchayati Raj institutions and co-operatives can play significant role in all these directions. Giving them power over the administration, as contemplated under the 73rd and 74th Amendment of the Constitution has not been implemented seriously so far in any of the states.

Disaster Management

The frequency and intensity of disasters such as floods, droughts, cyclones and earthquakes have increased in the recent years. The devastating earthquake in Gujarat has brought untold miseries to the whole state and caused a national disaster. Special effort should be made to develop appropriate technologies for increasing preparedness to predict and to manage the disasters. Effective and reliable information and communication systems, contingency planning and national and international mobilization of technologies and resources are a must. Experiences of other countries in prevention and management of the disasters should be shared.

Keeping Pace with Globalisation

The globalization of agricultural trade will bring to the fore access to markets; new opportunities for employment and income generation; productivity gains and increased flow of investments into sustainable agriculture and rural development. I believe that if managed well, the liberalization of agricultural markets will be beneficial to developing countries in the long run, It will force the adoption of new technologies, shift production functions upwards and attract new capital into the deprived sector. However, this will only come to pass if we are mindful of the interests of billions of small and subsistence oriented farmers, fisher folk and forest dwellers in the short and medium tern.

So far the magic of globalization has not been felt in India. During the past one-decade of liberalization certain trends such as deceleration of the growth rate of agricultural GDP, declaration in yield growth rates, and low non-agricultural employment have emerged against expectations. As we globalize, however, it is imperative that we do not forget social aspirations for a more just, equitable and sustainable way of life. Trade agreements must be accompanied by operationally effective measures to ease the adjustment process for a small farmer in developing countries.

Exploiting Cyberspace

Information is power and will underpin future progress and prosperity. Efforts must be made to strengthen the informatics in agriculture by developing new databases, linking databases with international databases and adding value to information to facilitate decision making at various levels. Development of production models for various agro ecological regimes to forecast the, production potential should assume greater importance. Using the remote sensing and GIS technologies, natural and other agricultural resource should be mapped at micro and macro levels and effectively used for land and water use planning as well as agricultural forecasting, market intelligence and e business, contingency planning- and prediction of disease and pest incidences.

ACCENT ON DIVERSIFICATION OF AGRICULTURE AND VALUE ADDITION

In the face of shrinking natural resources and ever increasing demand for larger food and agricultural production arising due to high population and income growths, agricultural intensification is the main course of future growth of agriculture in the region.

Research for product diversification should be yet another important area. Besides developing technologies for promoting intensification, the country must give greater attention to the development of technologies that will facilitate agricultural diversification particularly towards intensive production of fruits, vegetables, flowers and other high value crops that are expected to increase income growth and generate effective demand for food.

The per capita availability of arable land is quite low and declining over time. Diversification towards these high value and labour intensive commodities can provide adequate income and employment to the farmers dependent on small size of farms. Due importance should be given to quality and nutritional aspects. High attention should be given to develop post-harvest handling and agro-processing and value addition technologies not only to reduce the heavy post-harvest losses and also improve quality through proper storage, packaging, handling and transport. The role of biotechnology in post-harvest management and value addition deserves to be enhanced.

ACCENT ON POST-HARVEST MANAGEMENT, VALUE ADDITION AND COST-EFFECTIVENESS

Post-harvest losses generally range from 5 to 10 percent for non-perishables and about 30 percent for perishables.

This loss could be and must be minimized. Let us remember, a grain saved is a grain produced. Emphasis should therefore be placed to develop post-harvest handling, agro-processing and value-addition technologies not only to prevent the high losses, but also to improve quality through proper storage, packaging, handling and transport.

With the thrust on globalization and increasing competitiveness, this approach will improve the agricultural export contribution of India, which is proportionately extremely low. Cost-effectiveness in production and post-harvest handling through the application of latest technologies will be a necessity. The agro-processing facilities should preferably be located close to the points of production in rural areas, which will greatly promote off-farm employment. Such centres of processing and value addition will encourage production by masses against mass production in factories located in urban areas. Agricultural cooperatives and Gram Panchayats must play a leading role in this effort. In doing so, the needs of small farmers should be kept in mind.

Increased Investment in Agriculture and Infrastructures

The public investment in agriculture has been declining and is one of the main reasons behind the declining productivity and low capital formation in the agriculture sector. With the burden on productivity driven growth in the future, this worrisome trend must be reversed.

Private investment in agriculture has also been slow and must be stimulated through appropriate policies. Considering that nearly 70 percent of India still lives in villages, agricultural growth will continue to be the engine of broad based economic growth and development as well as of natural resource conservation, leave alone food security and poverty alleviation.

Accelerated investment are needed to facilitate agricultural and rural development through:

- Productivity increasing varieties of crops, breeds of livestock, strains of microbes and efficient packages of technologies, particularly those for land and water management, for obviating biotic, a biotic, socio economic and environmental constraints;
- Yield increasing and environmentally friendly production and post harvest and value addition technologies;
- Reliable and timely availability of quality inputs at reasonable prices, institutional and credit supports, especially for small and resource poor farmers, and support to land and water resources development;
- Effective and credible technology, procurement, assessment and transfer and extension system involving appropriate linkages and partnerships; again with an emphasis on reaching the small farmers;
- Improved institutional and credit support and increased rural employment opportunities, including those through creating agriculture based rural agro processing and agro industries, improved rural infrastructures, including access to information, and effective markets, farm to market roads and related infrastructure;
- Particular attention to the needs and participation of women farmers; and
- Primary education, health care, clean drinking water, safe sanitation, adequate nutrition, particularly for children (including through mid day meal at schools) and women.

The above investments will need to be supported through appropriate policies that do not discriminate against agriculture and the rural poor. Given the increasing role of small farmers in food security and poverty alleviation, development efforts must be geared to meet the needs and potential of such farmers through their active participation in the growth process.

Government should facilitate and support community level action by private voluntary organizations, including farmers groups aimed at improving food security, reducing poverty, and assuring sustainability in the management of natural resources. In addition, governments should enhance efforts to ensure good nutrition and access to sufficient food for all through primary health care and education for all.

Modern biotechnology tools, genetic engineering, as well as conventional breeding methods are all expected to play important roles in the generation of higher yielding, pest and stress resistant varieties of rice, wheat, maize and other cereal crops. The availability of genetic innovations in developing countries will depend on continued high levels of investments in agricultural research, both at the international and the national levels. Free and unhindered access to germplasm to breeders worldwide is absolutely crucial to the rapid dissemination and adoption of improved germplasm. This free movement and the dissemination of modern biotechnology innovations to developing countries are hampered by increased patent protection and private sector investments. There is an urgent need to address this problem of free access to technology in the future.

Increased attention will also have to be given to development of sustainable systems that protect the natural resource base. Recent evidence of resource degradation and declining productivity in some intensively cropped areas is of particular concern. Also population driven intensification of agriculture without the use of external inputs, is leading to a serious problem of mining soil fertility

Sustaining global food supplies will depend on continued high levels of investments in research and technology development. It is essential that research capacity has to be increased substantially. In addition to investments in research, infrastructure investments, particularly in irrigation, transport and market infrastructure development are equally important for sustaining the productivity and profitability of food crop production.

Mobilize the best of science and development efforts (including traditional knowledge and modern scientific approach) through partnerships involving national and international research institutions, NGOs, farmers' organizations and private sector in order to tackle the present and future problems of food security and production.

Donors and Government must urgently increase funding for agricultural research targeted at the needs of the rural and urban poor, and every effort must be made to ensure the free flow of information, technology and germplasm so that a proper sustainable agriculture can be achieved.

FIGHTING POVERTY AND HUNGER

Nearly one fourth of India's population, 251 million out of nearly one billion, is below the poverty line. One hundred seventy millions of the poor, 68 percent, are rural and the remaining 32 percent are urban. Number at the national level in rural area has decreased after 1983; the number of poor in the cities has been increasing. This is essentially due to migration of the destitute from villages to cities. There are serious implications of this trend on feeding the cities and food security of urban people, urban poverty and environment. A question may be asked as to whether the rural settings and opportunities could be improved for securing livelihood security and consequently rationalizing the migration to the cities.

An analysis of the incidence of rural poverty and hunger by farm size revealed that more than half of the landless people are poor. Poverty got significantly reduced from 54 percent in the landless group to 38 percent in the sub marginal group, suggesting that even a small piece of land, less than 1/2 hectare, can greatly reduce both poverty and hunger. The incidence of hunger and poverty gets reduced as one is able to meet even part of his/her dietary energy requirement through growing his/her own food. Studies show that even a small plot of one's own helps women to escape extreme poverty and deprivation. Land is the main asset for livelihood security.

Although several factors affect the extent and depth of poverty and hunger, some of them have overwhelming impacts under the Indian setting. These include, irrigation, farming system and literacy. Generally, there is higher concentration of poor, and hungry people in rainfed areas as compared with those in irrigated zones. Even with 20 percent of the irrigation intensity, there is a sharp fall in the proportion of hunger and poverty and it remains there irrespective of further intensification of irrigation. Evidences suggest that extensive irrigation will prove much more effective than to adding more and more water, and often wasting it along with the associated degradation of the natural resources. Such a policy will not only reduce poverty and hunger, but will also promote equity and environmental protection and natural resource conservation. An effective water policy and institutional support is needed to ensure judicious and equitable allocation, distribution and exploitation of water and water resources.

Livestock has the highest effect on reducing poverty and hunger. In rural India, 43 percent of the people who do not own even a single livestock are malnourished. Addition of one cattle or one buffalo to their assets reduces the hunger prevalence by 16 and 25 percentage points, respectively. Only 14 percent of the people who owned one cattle and one buffalo were malnourished. In urban areas also, the addition of one cattle or one buffalo had significant impact on reduction of proportion of malnourished people. Livestock sector should also receive high priority with multiple objectives of diversifying agriculture, raising income and meeting the nutritional security of the poor farm households.

Literacy has a very high impact on poverty alleviation as well as on hunger reduction. The illiterate people, whether urban or rural, are the most poor and malnourished. In urban areas the impact of literacy on poverty is the highest. Education, even above primary level, is extremely effective in reducing both poverty and hunger. Graduate and technical education is, of course, the most important instrument for reducing both poverty and hunger. But its impact is most visible on poverty reduction. Therefore, the education policy of the country must be geared to remove illiteracy as soon as possible, as 50 percent of our people are still illiterate. Free education up to 8'th standard coupled with mid day meal in the schools will go a long way in reducing both poverty and hunger and will thus help build a strong India. Further, this move will greatly reduce the violation of child labour laws and will offset some of the non tariff restrictions imposed by developed countries on exports from developing countries on the grounds of use of child labour.

INCENTIVES FOR AGRICULTURE

The Government will endeavour to create a favourable economic environment for increasing capital formation and farmer's own investments by removal of distortions in the incentive regime for agriculture, improving the terms of trade with manufacturing sectors and bringing about external and domestic market reforms backed by rationalization of domestic tax structure. It will seek to bestow on the agriculture sector in as many respects as possible benefits similar to those obtaining in the manufacturing sector, such as easy availability of credit and other inputs, and infrastructure facilities for development of agri-business industries and development of effective delivery systems and freed movement of agro produce.

Consequent upon dismantling of Quantitative Restrictions on imports as per WTO Agreement on Agriculture, commodity-wise strategies and arrangements for protecting the grower from adverse impact of undue price fluctuations in world markets and for promoting exports will be formulated. Apart from price competition, other aspects of marketing such as quality, choice, health and bio-safety will be promoted. Exports of horticultural produce and marine products will receive particular emphasis. A two-fold long term strategy of diversification of agricultural produce and value addition enabling the production system to respond to external environment and creating export demand for the commodities produced in the country will be evolved with a view to providing the farmers incremental income from export earnings. A favourable economic environment and supportive public management system will be created for promotion of agricultural exports. Quarantine, both of exports and imports, will be given particular attention so that Indian agriculture is protected from the ingress of exotic pests and diseases.

In order to protect the interest of farmers in context of removal of Quantitative Restrictions, continuous monitoring of international prices will be undertaken and appropriate tariffs protection will be provided. Import duties on manufactured commodities used in agriculture will be rationalized. The domestic agricultural market will be liberalized and all controls and regulations hindering increase in farmers' income will be reviewed and abolished to ensure that agriculturists receive prices commensurate with their efforts, investment. Restrictions on the movement of agricultural commodities throughout the country will be progressively dismantled.

The structure of taxes on foodgrains and other commercial crops will be reviewed and rationalized. Similarly, the excise duty on materials such as farm machinery and implements, fertilizers, etc., used as inputs in agricultural production, post harvest storage and processing will be reviewed. Appropriate measures will be adopted to ensure that agriculturists by and large remain outside the regulatory and tax collection systems. Farmers will be exempted from payment of capital gains tax on compulsory acquisition of agricultural land.

Investments in Agriculture

The agriculture sector has been starved of capital. There has been a decline in the public sector investment in the agriculture sector. Public investment for narrowing regional imbalances, accelerating development of supportive infrastructure for agriculture and rural development particularly rural connectivity will be stepped up. A time-bound strategy for rationalisation and transparent pricing of inputs will be formulated to encourage judicious input use and to generate resources for agriculture. Input subsidy reforms will be pursued as a combination of price and institutional reforms to cut down costs of these inputs for agriculture. Resource allocation regime will be reviewed with a view to rechannelizing the available resources from support measures towards assets formation in rural sector.

A conducive climate will be created through a favourable price and trade regime to promote farmers' own investments as also investments by industries producing inputs for agriculture and agro-based industries. Private sector investments in agriculture will also be encouraged more particularly in areas like agricultural research, human resource development, post-harvest management and marketing.

Rural electrification will be given a high priority as the prime mover for agricultural development. The quality and availability of electricity supply will be improved and the demand of the agriculture sector will be met adequately in a reliable and cost effective manner. The use of new and renewable sources of energy for irrigation and other agricultural purposes will also be encouraged.

Bridging the gap between irrigation potential created and utilized, completion of all on-going projects, restoration and modernization of irrigation infrastructure including drainage, evolving and implementing an integrated plan of augmentation and management of national water resources will receive special attention for augmenting the availability and use of irrigation water. Emphasis will be laid on development of marketing infrastructure and techniques of preservation, storage and transportation with a view to reducing post-harvest losses and ensuring a better return to the grower. The weekly periodic markets under the direct control of *Panchayat Raj* institutions will be upgraded and strengthened. Direct marketing and pledge financing will be promoted. Producers markets on the lines of Ryatu Bazars will be encouraged throughout the width and breadth of the country.

Storage facilities for different kinds of agricultural products will be created in the production areas or nearby places particularly in the rural areas so that the farmers can transport their produce to these places immediately after harvest in shortest possible time. The establishment of cold chains, provision of pre-cooling facilities to farmers as a service and cold storage in the terminal markets and improving the retail marketing arrangements in urban areas, will be given priority. Upgradation and dissemination of market intelligence will receive particular attention.

Setting up of agro-processing units in the producing areas to reduce wastage, especially of horticultural produce, increased value addition and creation of off-farm employment in rural areas will be encouraged. Collaboration between the producer cooperatives and the corporate sector will be encouraged to promote agro-processing industry. An interactive coupling between technology, economy, environment and society will be promoted for speedy development of food and agro-processing industries and building up a substantial base for production of value added agro-products for domestic and export markets with a strong emphasis on food safety and quality. The Small Farmers Agro Business Consortium (SFAC) will be energized to cater to the needs of farmer entrepreneurs and promote public and private investments in agri-business.

Institutional Structure

Indian agriculture is characterized by pre-dominance of small and marginal farmers. Institutional reforms will be so pursued as to channelize their energies for achieving greater productivity and production. The approach to rural development and land reforms will focus on the following areas:

- Consolidation of holdings all over the country on the pattern of north-western States;
- Redistribution of ceiling surplus lands and waste lands among the landless farmers, unemployed youth with initial start-up capital;
- Tenancy reforms to recognize the rights of the tenants and share croppers;
- Development of lease markets for increasing the size of holdings by making legal provisions for giving private lands on lease for cultivation and agri-business;
- Updating and improvement of land records, computerization and issue of land pass-books to the farmers, and
- Recognition of women's rights in land.

The rural poor will be increasingly involved in the implementation of land reforms with the help of Panchayati Raj Institutions, Voluntary Groups, Social Activists and Community Leaders.

Private sector participation will be promoted through contract farming and land leasing arrangements to allow accelerated technology transfer, capital inflow and assured markets for crop production, especially of oilseeds, cotton and horticultural crops.

Progressive institutionalization of rural and farm credit will be continued for providing timely and adequate credit to farmers. The rural credit institutions will be geared to promote savings, investments and risk management. Particular attention will be paid to removal of distortions in the priority sector lending by commercial banks for agriculture and rural sectors. Special measures will be taken for revamping of cooperatives to remove institutional and financial weaknesses and evolving simplified procedure for sanction and disbursement of agriculture credit. The endeavour will be to ensure distribution equity in the disbursement of credit. Micro-credit will be promoted as an effective tool for alleviating poverty. Self Help Group – Bank linkage system, suited to Indian rural sector, will be developed as a supplementary mechanism for bringing the rural poor into the formal banking system, thereby improving banks outreach and the credit flows to the poor in an effective and sustainable manner.

The basic support to agriculture has been provided by cooperative sector assiduously built over the years. The Government will provide active support for promotion of cooperative-form of enterprise and ensure greater autonomy and operational freedom to them to improve their functioning. The thrust will be on:

- Structural reforms for promoting greater efficiency and viability by freeing them from excessive bureaucratic control and political interference;
- Creation of infrastructure and human resource development;
- Improvement in financial viability and organizational sustainability of cooperatives;
- Democratisation of management and increased professionalism in their operations, and
- Creating a viable inter-face with other grass-root Organizations.

The Legislative and regulatory framework will be appropriately amended and strengthened to achieve these objectives.

SUSTAINABLE CREDIT FOR RURAL DEVELOPMENT

Few rural development endeavors confront the issues and challenges of sustainability more directly than the provision of credit to agricultural and rural producers. Agriculture lending, particularly to small farmers, has a nearly three-decade history of pairing donor agencies with developing country governments to promote economic development by alleviating credit constraints. Despite the relatively large amounts of credit directed towards rural producers, results have often been disappointing.

Provision of agricultural credit has been pursued in the context of both production and equity objectives. The strategy generally employed is to increase the total supply and reduce the cost. By making cheap credit widely available to the rural poor it was thought that production would be increased, with the added benefit of improving the income distribution of the rural poor.

Credit policies and programmes have been based on a set of "heroic" assumptions about small rural producers and their access to financial services. These include, for example, that small producers must be induced to adopt agricultural innovations by offers of cheap credit; that they have no savings capacity; and that with limited access to commercial credit they must rely on "usurious" informal sources, such as money lenders or pawnbrokers.

The lending landscape created by acting on these assumptions features convoluted regulations, special funds, complex rediscounting and reserve arrangements, loan guarantees, and other incentives to encourage lenders to increase credit flows to the rural poor. Special programmes and institutions, each with their own objectives, targeted particular groups to get credit, in what amount, and for which. Financial markets have become fragmented, rather than integrated, as new entities, such as agricultural credit agencies, development banks, and cooperatives were created to disburse and monitor these supervised credit programmes. With such a narrow focus on credit disbursement, many lenders either failed to, or were even prevented from providing deposit and savings services to rural producers.

Besides dysfunctional fragmentation, rural financial markets exhibit high degrees of inefficiency, ineffectiveness, and distortion. Populating the credit landscape are institutions with programmes plagued by loan recovery problems, outright failure to repay, chronic dependency on subsidies and external support, and excessive transactions costs for both the institutions and clients.

These institutions remain viable only due to continuous infusions of external resources. Further, they rarely reach their intended small farmer beneficiaries, and fail to achieve either their production goals, or any substantial increase in rural income.

Despite the history of failure, we cannot ignore the potential contribution that effective financial markets and financial institutions can make to economic development by stimulating savings and facilitating intermediation between borrowers and savers. There is a strong development rationale for public intervention to build up the formal (and informal) financial sector and extend services to groups that are not yet being adequately served. However, to be of ongoing benefit to developing countries, these institutions must prove sustainable over the long term.

Because the "good" that credit institutions deal with is monetized, relatively precise calculations of the cost and value of the benefits provided can be made, relationships between benefits and incentives can be quantified, and measures for sustainability developed. SCOPE highlights stakeholders and the value they attach to institutional outputs, factors that play key roles in rural credit. The framework also focuses attention on the complex and multiple exchanges between institutions and actors in the environment.

This chapter examines the credit component of the USAID-funded Provincial Area Development Programme (PDP) in Indonesia using the SCOPE concepts. Credit systems are reviewed in three provinces: Central Java, South Kalimantan, and West Nusa Tenggara.

The case study focuses on the development of capacity and performance of core insti tutions involved in the various PDP credit systems; the impact of the provincial credit systems on borrowers and other beneficiaries; and the credit systems' viability and prospects for sustainability. The study seeks to convey a flavour of the complex interactions among institutional actors that played an important role in shaping the development of the credit system.

RESEARCH ON FARM STRUCTURE AND RURAL COMMUNITIES

A further area of dissatisfaction with research on farm structure and rural communities is that the treatment of farm structure in this literature has generally failed to incorporate new concepts relating to agricultural dualism. Typically farm size or scale is operationalized as a single variable representing a central tendency measure of farm size, which was a suitable procedure when there was a singular process of differentiation among farms. But with the emerging dualism in U.S. farm structure since the late 1970s, there has been little change in mean acres per farm since the 1969 census. This measure of central tendency thus ignores the processes of dualism that have contributed to increased concentration with apparent stability in average acres per farm since the late 1960s.

The neo- Goldschmidt conceptualization of an agricultural structure in which particular communities can be characterized along a continuum of farm sizes ranging from small to large has also been questioned by Wimberley who, using multiple indicators from county-level Census of Agriculture data and a longitudinal design, found that there are instead three continua relating to degree of "corporate-commercial" agriculture, "largefarm-area" agriculture, and "small-farm" agriculture. Using Wimberley's measures and including sources of off-farm employment as well, Reif found that large-farm-area agriculture and core-industry structures were related both to family poverty and income inequality in counties.

Thus, while studies of the impacts of changing farm structures on rural communities have been a productive area of research in the new sociology agriculture, a number of research frontiers for the future are suggested by recent criticism. There is a need to diversify the methodological approach, consider alternative directions of causality, introduce more sophistication in terms of the measurement of change in farm structure, and expand recent efforts at comparative regional and interdisciplinary research.

The work of Vail represents an encouraging thrust in this direction. Vail has conducted extensive research on the relationships between agriculture and rural communities in Maine in which a variety of factors of demonstrated importance — the reciprocal relations between agriculture and rural communities, the role of an increasingly dualistic agrarian structure, nonfarm aspects of agriculture, and the historically specific nature of farm-community interactions — have been incorporated into a single study.

THE FARM CRISIS OF THE 1980S

The notion that American agriculture is no longer (if it ever was) isolated from the remainder of the political economy was driven home clearly by the farm crisis of the 1980s. This crisis was brought by the conjunction of a number of factors in the 1970s, including: accelerated demand for grain on world markets, public policies that encouraged massive capital investments and expansion of production, and a political-economic context that encouraged land speculation. The farm crisis emerged in the 1980s for a number of reasons. Global agricultural export markets shrank, in large part due to the global recession. Farm commodity prices and ultimately net farm incomes (exclusive of federal commodity programme payments and price supports) declined due to the glutting of world grain markets. Reagan administration fiscal and monetary policies led to high real interest rates, which were especially detrimental to agriculture because of its interest rate sensitivity; accordingly, debt-asset ratios among farmers increased rapidly.

The 1980s also witnessed the emergence of new Third World agricultural export competitors (*e.g.*, Argentina, Brazil) because of the global debt crisis and these countries' need to increase exports in order to repay foreign loans. Further, federal fiscal austerity eroded the federal government's ability to maintain farm incomes through commodity programs (even though commodity programme expenditures soared during the decade). The result was a massive decapitalization of American agriculture, especially in the Midwest and Great Plains, as farmland prices in the states most affected by the crisis declined by over 50 percent.

This crisis attracted a good deal of attention by the mass media and by agricultural economists. It was not, however, until the middle of the decade that much was published about the farm crisis by rural sociologists. Most of this research can be divided into four categories: (1) neo-Marxian (and related neo-Weberian) analyses of the continuing penetration of capitalism into and the role of the state in agriculture, (2) analyses of the differences in the characteristics between farm(er)s that were and were not experiencing "financial stress," (3) studies of the effects of farm financial crisis and failure on the farm families whose enterprises were affected, and (4) studies of public reactions to the farm crisis.

The neo-Marxian (and neo-Weberian) approaches have largely focused on understanding continuing cycles of crises in agriculture under capitalism. Although these analyses focus heavily on the political and economic context of agriculture, they are not fundamentally different from what is discussed and therefore are not discussed specifically here. Examples of this genre specifically focused on the crisis of the 1980s include Berlan, Bonanno, Mooney, McIntosh and Zey-Ferrell, Zey-Ferrell and McIntosh, Goodman and Redclift, and Buttel. We also briefly examine the farm crisis from a methodological angle — namely, the growing tendency in the political economy of agriculture to stress periods of agricultural crisis as key "test cases" for assessing theories of agrarian transition.

Studies in the second category, analyses of the characteristics of farms with various levels of "farm financial stress," deal more with the effects of the political and economic environment of agriculture on farms of different characteristics. These studies have produced inconsistent results due to regional, ethnic, and other variations within the farming population.

For example, farm-size variables were found to be directly correlated with financial difficulty on Iowa farms and unrelated for Texas and North Dakota farms. Yet aggregate national data revealed that larger farms were more profitable, a relationship that obscured the underlying failure of many middle size farms. Farm families with fewer years of farming and younger ages were affected more severely.

However, Murdock et al. found the cohort that had started farming between the end of World War II and the mid1960s also had higher rates of financial difficulties, suggesting a life-cycle effect for those farm families bringing another generation into the farm. The literature does, however, show agreement on the relationship between other variables and financial crisis. For example, higher debt-to-asset ratios brought by low commodity prices in a time of deflation of asset values were associated with financial problems.

The third category, analysis of the effects of the farm financial crisis and failure on the farm families whose enterprises were affected directly, has largely been the work of William and Judith Heffernan, who have conducted in-depth interviews among Missouri farm families who were in severe financial crisis or had been displaced from agriculture. These families experienced a host of difficulties beyond the simple loss of their farms and the disorientation that accompanies the loss of a valued way of life. Among their major results are the following: Many farm families felt socially isolated and were not well-prepared for participation in nonfarm labor markets. Local employment opportunities were often unavailable, and tax laws devastated some families economically after they had attempted to leave farming with some resources with which to start a new life. While the long-term effects of farm failure for these families were not yet known, it was clear that some families were adapting to their new situations successfully while others were at risk of suffering long-term poverty. Graham's study of displaced farm families in New York State bears out these findings. This category of studies has implications for the mutual effects of agriculture and its environment as the displaced farm families affect local labor markets and community social structures.

The final category of research on the 1980s farm crisis, on public reactions to the crisis, is best represented by Lyson's research on public apathy. Drawing on theories of public opinion formation, Lyson demonstrates that several factors led to a surprisingly high degree of apathy by nonfarm people towards the farm crisis, despite the massive media attention the crisis attracted. He notes that there has been a trend towards farmers and consumers having very little direct contact and that the relationship between food prices and farm income has become concealed. Most importantly, Americans have come to view "bad news" from the farm sector — the cost-price squeeze, high debt loads, and the loss of family farms — as being normal. Thus the American public, though "interested and concerned" about the economic woes of farmers, exhibited apathy towards the farm crisis.

A major neglected area in the rural sociology literature on the farm crisis to date is studies of the effects of the farm crisis on rural communities. This topic pertains, in particular, to how conditions in agriculture affect its environment, especially at the local level. Financial losses to local agribusiness firms when farms fail, losses of farm families in local rural market areas, declining property tax revenues due to deflated farm asset values, and other problems associated with the declining prosperity of agriculture have likely had severe effects on local services and on the local social fabric.

AGRICULTURE AND THE RURAL COMMUNITY

Perhaps in no other area of the "new sociology of agriculture" is there so much apparent continuity with the pre-Korean War tradition of rural sociological research as in research on farm structure and rural communities. The early rural sociologists primarily conceived of agriculture as one of the major institutional ensembles affecting rural community life and, especially during the 1930s, devoted much of their research effort relating to agriculture to understanding the implications of agricultural structural changes for rural community life and viability.

The conceptualization of agriculture and rural community interactions and the methodologies for exploring these interactions in pre-Korean War and post-1970 rural sociological research, however, exhibit dramatic differences. Whereas most of the early rural sociologists tended to be less interested in agriculture per se and were principally concerned with understanding rural community institutions, contemporary researchers have tended to be more interested in structural change in agriculture and have deemphasized understanding the totality of the fabric of rural social life.

The research tools and data sources of these two generations of rural sociological researchers are also dramatically different. The early rural sociologists relied on rural social surveys, while the bulk of contemporary research is based on county-level census data. A further difference between the early and contemporary rural sociologists of agriculture in their work relating to rural communities concerns the pivot of the latter studies around Goldschmidt's "As You Sow". This is somewhat ironic in that Goldschmidt's study was anthropological and qualitative, whereas the research conducted since 1970 that has drawn on Goldschmidt's work has largely been quantitative and has involved far less "anthropological" detail on rural communities than did Goldschmidt's.

Rodefeld is generally credited with the "rediscovery" and popularization of Goldschmidt's pioneering work among rural sociologists, though Heffernan (1972) made important contributions to elevating "Goldschmidt-style" research to a prominent position in rural sociology. Nonetheless, after Rodefeld's refocusing of rural sociological attention on the Goldschmidt thesis, Goldschmidt's book has been virtually an obligatory citation in research articles on this topic. Despite some trenchant criticisms of Goldschmidt's perspective and research methodology, Goldschmidt's general hypotheses remain accepted theoretically and thus continue to shape, if not dominate, research on farm structure and rural communities within the "new sociology of agriculture."

Briefly, Goldschmidt's thesis was that there are differences in the quality of rural community life depending upon the social organization of agriculture and its attendant occupational structure. Goldschmidt observed that the quality of rural life in two carefully matched California communities in the San Joaquin Valley differed dramatically and that these differences were largely accounted for by the organization of agriculture.

He argued that Dinuba, a community surrounded by "family" farms, had a greater number of businesses, a larger retail sales volume, a higher per capita income, and a greater diversity of social, educational, recreational, and cultural institutions than did Arvin, a community surrounded by large, industrial, corporate farms. These differences were principally accounted for by an intervening variable — community occupational composition — that was seen to be strongly influenced by the social organization of agriculture and which directly contributed to the dramatic variations in indicators of the quality of life. Goldschmidt reported that the high ratio of laborers to other occupational groups in Arvin accounted for its low level of community viability relative to Dinuba. From these observations Goldschmidt inferred that the rise of large-scale corporate agriculture then underway in California and elsewhere would have adverse implications for rural communities.

Goldschmidt's study received essentially no further attention until it was referred to in a 1968 hearing on corporation farming before the Senate Select Committee on Small Business. LaRose and Sonka and Heady soon reported further data consistent with the Goldschmidt study. In addition to Rodefeld's and Heffernan's efforts to resurrect Goldschmidt's work in rural sociological circles, there was a major research effort by the Small Farm Viability Project in California, drawing largely on Isao Fujimoto's unpublished research, which presented a larger restudy of the Goldschmidt thesis. The results largely supported Goldschmidt's findings of over three decades earlier. Flora and Conboy also reported results consistent with the thesis.

Since that time there have been numerous studies in the new sociology of agriculture tradition that have explored the Goldschmidt thesis in a number of contexts. Several observations can be made on the conduct of recent research in the Goldschmidt tradition, what we might call "neoGoldschmidt" research. First, it must be stressed that contemporary rural sociological researchers have taken considerable liberties with Goldschmidt's formulations. As noted earlier, Goldschmidt's method was a comparative anthropological case study of two communities and eschewed quantitative analysis of the sort that has predominated in the subsequent rural sociological literature.

Also, Goldschmidt's analysis focused on several aspects of the property relations and control over labor, such as absentee ownership and use of ethnic minority laborers, rather than farm size, which has been the most important farm structural indicator in the neo-Goldschmidt tradition. In that regard it is important to stress that the "family" farms of Dinuba were by no means "small" farms; in fact, the family farms of Dinuba were generally larger in gross sales terms than most large farms in the Midwest today.

Second, neo- Goldschmidt research has largely focused on a single direction of causality — the impact of agricultural structure on rural communities — and has largely ignored other types of interrelations between agriculture and rural communities. Third, most studies have been based on Census of Agriculture data and have employed countylevel data as a proxy for community characteristics have utilized survey data to infer community quality of life consequences of changing agricultural structures. Fourth, despite obvious major regional variations in farm structures and in rural communities and their economies, there has, until very recently, been a virtual lack of a comparative perspective on farm and community structures.

Goldschmidt and Harris and Gilbert are partial exceptions to this generalization, since they have done research using states as the unit of analysis, though this procedure compounds the aggregation bias problems. Very recently, nonetheless, there has appeared a series of regionally based studies of the effects of farm structure on rural community viability and quality of life commissioned by the Office of Technology Assessment. These studies are Buttel et al., Skees and Swanson, Flora and Flora, MacCannell, and van Es et al. as published in abridged form in Swanson. These five studies were very briefly summarized in the main report. Together they indicate strongly that there are pronounced regional variations in how changing farm structures affect rural communities.

The impacts of farm structural changes on rural communities are generally most pronounced in the Great Plains and industrial agriculture states, less so in the South and Midwest, and very little in the Northeast. There is evidence that the degree to which farm structure affects rural communities is directly associated with the dependence of the rural community on agriculture.

The general model pursued in these neo- Goldschmidt studies has been to estimate coefficients for a causal chain in which farm structural change indicators are seen to affect the composition of the farm population (*e.g.*, as measured by indicators of the size of the farm population, the proportion of full-time hired workers) and the size and composition of the rural population. These clusters of variables are then seen to affect indicators of the quality of life in rural-agricultural communities (*e.g.*, proportion of families in poverty, median family income, per capital retail sales, unemployment rate).

This general model is sometimes supplemented by technological factors as antecedents of farm structure, since technological change is considered to be one of the major factors influencing farm structural change. More recently, studies in this genre have gone beyond the cross-sectional census data designs that predominated from the mid-1970s to the early 1980s and have included measures of change, where appropriate, in variables in the model. One limit to studies that use census and other data based on single-year assessments of farm income is that net farm incomes are highly variable over the short-run; for example, as many as 20 percent of farms with annual sales of over $250,000 had poverty-level incomes in 1986.

Heffernan has effectively summarized the general configuration of results in the neo-Goldschmidt literature by noting that "all relevant research to date suggests that a corporate type of agriculture results in a reduction in the quality of community life for at least some people, especially the hired workers in rural communities." As such, the results continue to lend support to the Goldschmidt thesis. But there have recently been several expressions of dissatisfaction with Goldschmidt's perspective and with the generalizations that have been made from research in this tradition.

A major theoretical concern is that Goldschmidt and many researchers working in this tradition have taken large-scale and corporate agriculture to be basically synonymous when, in fact, not all large-scale farms are corporate or industrial-type farms. Much more important than industrial-type farms in farm numbers and shares of sales, assets, and profits are larger-than-family farms. Empirically, as Goldschmidt-style research has been done in the Midwest and other traditional family farming areas, there have been admonitions that large-scale agriculture, especially of a nonindustrial character, does not necessarily lead to adverse impacts on rural communities, though some studies note that increased scale of production could have detrimental impacts on community services.

THE PROVINCIAL AREA DEVELOPMENT PROGRAMME

Indonesia's PDP was implemented over a ten-year period, beginning in 1978 in two provinces, one of which was Central Java. It expanded to six, including South Kalimantan, in 1979, and encompassed eight by 1980, reaching West Nusa Tenggara in that year. The project design provided resources and support for the planning, implementation, monitoring, and evaluation of small subprojects, to increase the incomes of the rural poor by decentralizing development efforts to the provincial level and below. In the eight provinces, PDP undertook a wide range of sectoral activities, including agricultural food crops, estate crops, livestock, fisheries, village industry and handicrafts, manpower training, social welfare, and cooperative development.

PDP's objectives included improving

1. The capability of local governments to undertake rural development activities that improve the productive capacity of the rural poor;
2. The capability of the central government to support local governments in planning, implementing, and evaluating activities that improve the productive capacity of the rural poor; and
3. The incomes of the rural poor within the project areas through implementing small subproject activities. These objectives shifted over time, especially after 1982. They gradually moved away from experimental, direct, beneficiary-oriented subprojects, which were expected to be taken up by local government agencies, towards the development of institutional capability and introduction of planning and monitoring systems.

The project worked at the central level with the national planning agency, and at the provincial level, beginning in 1980, with decentralized regional planning boards (BAPPEDAs). PDP also worked through local sectoral agencies (*dinas*) at the provincial level and at the rural district (*kabupaten*) and sub-district (*kecamatan*) levels with local government entities.

From the outset, participating provinces were required to make some arrangement for the provision of credit in support of the various sectoral development efforts. How credit was structured and provided, varied considerably from province to province. Initially, the dinases handled the details of the credit programme through individual subprojects, but this proved to be a disaster. The local sectoral agencies had little difficulty in supplying credit as input to their production efforts, but had little capacity to manage repayment. Demand for credit in PDP came from the sectoral agencies to support their production efforts in the many subprojects that were being sponsored. Stimulating commodity production was the main focus of effort, and getting funds out to producers in support of the subprojects was the major emphasis of credit. Since lending is only one element of the activities a viable credit system needs to engage in, financial services delivery to beneficiaries began to founder. Repayment, covering operating costs, and savings mobilization were all neglected.

One notable exception to the pattern of failure was the sub-district credit agency (badan kredit kecamatan), or BKK, in Central Java province, supervised by the Regional Development Bank (BPD) and supported by PDP. Here the emphasis was on institutional development over strict credit delivery.

The success of the BPD's Central Java BKK operations during the early years of PDP attracted considerable attention, and led to emulation in other provinces.

As a result, the BPD was assigned the responsibility for providing PDP credit, including credit management, training and supervision, through the BKKs' networks of local offices. In conformity with the PDP objective of targeting the rural poor, credit operations were directed at small enterprises that were not served by formal lending institutions.

The three provincial BKK systems reviewed subsequently, illustrate PDP efforts directed at creating and maintaining viable institutions capable of responding to the demand for financial services in rural Indonesia. The three credit systems vary with respect to size, scope of operations, facilities, staffing, management, and leadership. In part, this variation reflects different strategies being pursued in the delivery of financial services, but it is also a function of the different levels of development of these systems.

THE CENTRAL JAVA BKK

In 1970, the BKK system was created as a project of the Central Java provincial government, which capitalized each individual local unit. By 1979, about one-third of the units had ceased to exist or were operating at such low levels as to make them non-viable. PDP provided support to rehabilitate the BKK. Interventions began in 65 units in the five rural sub-districts designated as PDP's target area. The institution-building experiment consisted of:

1. Recruitment and training or retraining of staff;
2. Development of accounting, supervision and auditing, reporting and classification procedures for the system as a whole; and
3. Increased capitalization on a graduating scale based upon unit classification.

The intent was to deal with immediate performance problems and build future capacity as well. Within the first three years of the experiment, substantial progress was made in improving the operations of targeted units. The success attracted the attention of the central government, which in 1981 provided a loan of $4.7 million to extend the rehabilitation efforts through the remainder of the province's sub-districts.

SIZE AND SCOPE OF OPERATIONS

The Central Java BKK has units in 497 sub-districts, and operates over 3,000 mobile posts that follow village market day schedules. It serves around half a million clients, and continues to grow. The BKK is primarily a lending institution, but targets savings mobilization as well. Compulsory savings is a loan requirement designed to inculcate new behaviours and to increase available loan capital. Following a change in regulations, the BKK now offers savings accounts to non-borrowers and is aggressively pursuing this new market. Average loan size is $36, with the borrowing period averaging 12 weeks.

FACILITIES AND STAFF

For the first ten years of its existence, the BKK operated out of borrowed rooms in the kecamatan government complex with borrowed or nonexistent equipment and material. Units were staffed with local government employees. The PDP institution-building intervention led to changes. The project established a revolving fund, managed by the BPD that BKK units meeting profitability criteria could borrow from, for equipment.

Permanent offices were financed the same way, using funds from the provincial development budget. This practice limited public subsidies and put government resource allocation on a profitability basis. In the mid- 1980s, the BKK shifted its staff from civil servant status to employees of the system itself, which allowed the introduction of performance-based pay, profit-sharing incentives, and more flexibility to hire and fire. Each unit has three core staff, a manager, cashier, and bookkeeper; hiring additional staff depends on service demand and profitability.

LEADERSHIP AND MANAGEMENT

Key actors providing leadership for the BKK have been the director of the BPD, and an assistant to the provincial government secretary. They maintained linkages with the governor's office and the BAPPEDA, lobbying and negotiating for policies favourable to the BKK.

The system also has a supervisory board that provides both leadership and oversight, and reports to the district head. The BKK's official designation as a sub-district institution has given the sub-district head a monitoring role, which has had a positive effect on performance. In addition, village chiefs serve as guarantors of the reliability of individual borrowers and as information conduits regarding credit matters.

THE BKK IN SOUTH KALIMANTAN

The BKK in South Kalimantan is not only a direct product of PDP credit interventions in the province, but as its name implies, is also a conscious replica of the Central Java system. Established in 1985, the South Kalimantan BKK has deliberately limited its services to the provision of small, short-term loans. The BKK serves 28 sub-districts of the 106 in the province. Loan packages in South Kalimantan tend to be larger and about twice as long in duration as those in Central Java. A decision not to offer a savings programme at the outset was based on considerations for the difficulty in training new staff to perform complex bookkeeping functions for two separate accounts simultaneously. The BPD has stated that it is making plans for the addition of a compulsory savings programme.

Facilities for the BKK, including equipment and motorcycles, were financed by the district with PDP funds, which are not subject to repayment by the system. PDP provided another subsidy in the form of two years' worth of staff and operating costs for each BKK unit. Staffs are employees of the credit agency, with three per unit.

As in Central Java, the BPD director has played a lead role in maintaining close contact with the provincial development planning board and the province governor on BKK matters. The system does not have a high-level supervisory board, though it does involve the sub-district head in performance monitoring.

TECHNOLOGICAL NEEDS AND FUTURE AGRICULTURE

It is apparent that the tasks of meeting the consumption needs of the projected population are going to be more difficult given the higher productivity base than in 1960s. There is also a growing realization that previous strategies of generating and promoting technologies have contributed to serious and widespread problems of environmental and natural resource degradation. This implies that in future the technologies that are developed and promoted must result not only in increased productivity level but also ensure that the quality of natural resource base is preserved and enhanced. In short, they lead to sustainable improvements in agricultural production.

Productivity gains during the 'Green Revolution' era were largely confined to relatively well endowed areas. Given the wide range of agroecological setting and producers, Indian agriculture is faced with a great diversity of needs, opportunities and prospects. Future growth needs to be more rapid, more widely distributed and better targeted. Responding to these challenges will call for more efficient and sustainable use of increasingly scarce land water and germplasm resources. Technical solutions required to solve problems will be increasingly location-specific and matched to the huge agroecological/climatic diversity. Detailed indigenous knowledge and greater skills in blending modern and traditional technologies to enhance productive efficiency will be more than ever before, key to the farming success and sectoral growth. Most technological solutions will have to be generated and adapted locally to make them compatible with socio-economic conditions of farming community.

New technologies are needed to push the yield frontiers further, utilize inputs more efficiently and diversify to more sustainable and higher value cropping patterns. These are all knowledge intensive technologies that require both a strong research and extension system and skilled farmers but also a reinvigorated interface where the emphasis is on mutual exchange of information bringing advantages to all. At the same time potential of less favoured areas must be better exploited to meet the targets of growth and poverty alleviation.

These challenges have profound implications for products of agricultural research. The way they are transferred to the farmers and indeed the way research is organized and conducted. One thing is, however, clear – the new generation of technologies will have to be much more site specific, based on high quality science and a heightened opportunity for end user participation in the identification of targets. These must be not only aimed at increasing farmers' technical knowledge and understanding of science based agriculture but also taking advantage of opportunities for full integration with indigenous knowledge. It will also need to take on the challenges of incorporating the socio-economic context and role of markets.

With the passage of time and accelerated by macro-economic reforms undertaken in recent years, the Institutional arrangements as well as the mode of functions of bodies responsible for providing technical underpinning to agricultural growth are proving increasingly inadequate. Changes are needed urgently to respond to new demands for agricultural technologies from several directions. Increasing pressure to maintain and enhance the integrity of degrading natural resources, changes in demands and opportunities arising from economic liberalization, unprecedented opportunities arising from advances in biotechnology, information revolution and most importantly the need and urgency to reach the poor and disadvantaged who have been by passed by the green revolution technologies.

Another important implication of increasing globalization relates to the need for greater attention to the quality of produce and products both for the domestic and the foreign markets. This would imply that production must be tuned to actual rapidly changing product demand. Such adaptation to global markets would require state of the art research, which can be achieved only by setting global standards of research, focus on well defined priorities and mechanisms which permit close interaction of farmers with researchers, the private sector and markets.

INDIA'S POSITION IN WORLD AGRICULTURAL TRADE: AN OVERVIEW

Global agricultural exports (including food products) increased from USD 326 billion in 1990 to USD 520 billion in 2003. Strong expansion in food and agri exports in the mid 1990s, was followed by a decline from 1997 to 2000.This was mainly due to a slump in prices for various agricultural commodities, on the back of demand-supply mismatches. Since 2001,food and agri trade has been growing.

The main items in food and agri trade include fruits and vegetables (-USD 75 billion),cereals and cereal-based preparations (-USD 58 billion), marine products (-USD 57 billion), meat and meat-based preparations (-USD 46 billion), milk and milk products (-USD 33 bn) and beverages (-USD 40 billion). Latin America has by far the largest agricultural trade surplus, followed by Oceania.

Brazil and Argentina contribute significantly to the Latin American trade surplus. Asia has the largest trade deficit of all regions, primarily on account of Japan. Western and Eastern Europe as well as Africa have a trade deficit too. About half of global trade flows are regionally contained, partly as a result of regional agreements and partly due to the perishable nature of agri-commodities.

There has been a sharp increase in regional trade agreements (RTAs) over the past decade. A total of 259 RTAs have been notified to the WTO by the end of 2002, although only 176 RTAs are operational. An additional 70 RTAs are estimated to be operational, but not yet notified by WTO and about 70 are under negotiation. A well known example of regional trade is in the European Union, where about 75 per cent of agricultural trade is within the region. The intra-regional trade figures for other regions are 33 per cent (NAFTA),63 per cent (Asia) and 15 per cent (Latin America and Africa). The group of G-20 countries, a grouping of developing countries led by Brazil, China, India and South Africa have sought further policy reforms and market liberalisation on the part of the developed world. At the Geneva meeting of the WTO, the members have agreed on a further reduction of overall trade distorting domestic support, the elimination of all forms of export subsidies and a reduction of import tariffs.

The EU already anticipated this by drastically reforming the Common Agricultural Policy (CAP). Besides agricultural policies and trade arrangements, demand-supply trends and the macroeconomic environment also significantly influence trade. Another important factor is the change in currency rates.

EXPORTS

India's exports of agricultural and food products are about INR 360 billion (approx. USD 8 billion) which constitutes about 1.6 per cent of total global trade

(USD 520 billion). Exports of agriculture and food products have grown at 15 per cent annually, vis a vis total exports which have grown at 19 per cent, in the last decade. The share of agricultural exports as a percentage of India's total exports has decreased from 19 per cent to 13 per cent in this period. Excluding non-food agricultural products (such as paper, cotton and jute), tobacco and seeds; marine exports constitute the single largest product category in agricultural and food exports from India followed by, rice oilmeals, wheat, cashew, tea and coffee.

Most of the agricultural products are primary processed. Many products have shown negative or single digit growth. The top 15 export categories in food and agricultural (except for non-food items and tobacco) from India. With the exception of buffalo meat, rice and cashew, India's share in global agricultural and food trade is insignificant. Marine products which have the highest share in Indian exports, have a 2 per cent share of global trade in this category. Among the key food exports, only non-basmati rice, wheat and buffalo meat has shown double-digit growth. Marine products, basmati rice and processed fruit juices/vegetables have shown single digit growth. Products like cashew, tea, spices, coffee have shown negative growth.

IMPORTS

Against exports of approximately USD 8 bn India's import of agricultural products are about USD 5 bn which translates into trade surplus of approximately USD 3 bn (Year 2003-04). India has abolished restrictions on imports of many agricultural and food products post liberalisation. However, tariff quotas are maintained on some edible oils, maize and milk powder.

There are restrictions on imports of certain fats, oils of animal origin, and beef, based on GATT Article XX that permits countries to restrict imports on religious grounds. Some products, such as wheat, rye, oats, maize, rice, canary seed and other cereals, continue to be traded by the Food Corporation of India. Despite liberalisation of imports, India's imports are growing at about 1 per cent per annum for the last five years. While exports cover a wide range of products, imports are skewed towards pulses and oilseeds.

Pulses and oilseeds are the key food items in India's imports (73 per cent share) of food items displaying year on year growth. Increasing production of edible oils in India can reduce India's dependence on imports to a large extent. Another category which has witnessed growth in imports is Scotch whisky. The volume growth is about 13 per cent and value growth is about 6 per cent in the last decade.

However, the growth is driven by imports of bulk Scotch, which is bottled in India and/or blended with Indian liquor by domestic manufacturers. Bottled Scotch imports have not witnessed any growth due to the current level of import duties.

ISSUES HINDERING EXPORTS

India continues to be either absent or at best a marginal player in most of the leading markets for its exports. Indian players have not succeeded in establishing direct linkages with buyers/consumers in importing countries, as a result of which a large proportion of exports are being further processed and re-exported by other countries. The sector-specific issues for key products arrived on the basis of discussions with leading exporters from India are highlighted below.

Product: Marine products - Decline in raw material availability:

- Anti-dumping issues (such as excess use of antibiotics) in key markets.
- Fragmented base of suppliers.
- High duties on imports of additives/flavourings for making value added products.
- Lack of availability of technology, *e.g.,* for tuna fishing.

Rice:

- Low cost competitiveness due to state taxes and MSP regime (for non-basmati]
- Outdated milling technology (using rubber roll sheller) resulting in high share of broken rice.
- Poor quality of seeds/non-availability of certified authentic seeds leading to lack of consistency in grain quality.

Cashew:

- Lack of established quality parameters for raw cashew.
- Lack of mechanisation of processing leading to higher production costs.
- Lack of raw material availability - high dependence on imports of raw nuts, Interstate barriers.
- On movement of raw nuts.
- Purchase tax for exports in some states (*e.g.*4 per cent in Tamil Nadu)

Product: *Key issues hindering exports*:

- Coffee- Low realisations due to exports restricted to green coffee and no exports in roasted/instant/branded coffee.
- Export cess of INR 500/MT leading to low profitability for exporters.
- High cost of production.
- Production focused on CTC rather than orthodox, whereas demand for orthodox tea is growing faster than for CTC.
- Tea- Export market viewed as a 'residual market' to sell surplus production.

Buffalo Meat

Mango:

- Absence of policy on permitting rearing for slaughter which translates into lack of traceability a key requi-rement in several importing countries.
- Inappropriate facilities at existing municipal slaughter houses.
- Lack of exportable varieties (high fibre content; inappropriate appearance and texture and large size of stone).
- Lack of packhouses from farm to port.
- Lack of post-harvest treatment facilities such as for vapor treatment.
- License for setting up private slaughter house is difficult to obtain.
- Prevalence of various diseases particularly Foot and Mouth Disease (FMD).

Grapes:

- High cost for obtaining certification for exports. For, *e.g.*, the cost of EurepGap certificate is.
- High cost of setting up vineyards.
- INR. 75000/farmer (including the cost of construction of separate storage space for fertilizers and pesticides etc).

Dairy Products:

- Few local manufacturing facilities of value added products.
- Lack of appropriate packaging technology.
- Lack of quality monitoring mechanism in the supply chain as required by many importing countries.
- Opportunistic production of SMP with low exportable surplus.

Guar gum:

- Increasing domestic demand from non-food applications such as textiles, paper, pharmaceuticals, oil well drilling.
- Poultry Products - Lack of competitiveness due to high cost of production.
- Variation in production from year to year (high share of rain-fed areas).

Uneconomic scale of operations:

- Inadequate and inappropriate storage and distri-bution infrastructure.
- Lack of consistency in supply and quality.
- Lack of cost competitiveness due to statutory charges, intermediation and wastages/losses.

Non- Tariff Barriers Short Product Life Cycle Lack of Brand Image

Most exporters from India lack scale- for example the largest fresh produce exporter records annual sales of about INR500mn. The low volume translates into lack of economies in operations and makes exports uncompetitive. Hence, exporters are not able to establish themselves as long-term players in the export market, and rely heavily on opportu-nistic businesses.

These factors cumulatively translate into low investments in upgrading skill sets, product innovation, quality improvement and brand building.

- Countries across the world have successfully addressed these issues to become globally compe-titive.
- Low competitiveness in global markets unable to establish as a long-term player.
- Low economies of scale high cost and low volume.
- Low efficiencies technological and managerial.
- Low investment capacity in product innovation and Brand/Market development.

Rice, Cashew and Coffee Exports from Vietnam

Cashew - Mechanisation of processing technology: Vietnam is the largest cashew producer in South East Asia and the third largest cashew exporter in the world after India and Brazil. Over 90 per cent of production is exported. The number of cashew exporters have increased from 16 in 1997, to over 50 in 2003. Coffee - Government Support: Vietnam is ranked as the second largest coffee exporter in the world since 2000, next to Brazil. It exports coffee to about 62 countries. The two largest markets include the European Union (47 per cent share) and the USA (15 per cent share).The Government has played an important role in increasing coffee exports. Some of the initiatives include: Cropping patterns have been adjusted to increase land area for planting winter-spring paddy and summer-autumn paddy (from 2.1 and 1.2 million ha in 1990 to 2.89 and 2.35 million ha in 2000 respectively) and reducing acreage under low - yielding winter paddy (from 2.74 to 2.4 million ha).

In order to cope with the market changes, the Gove-rnment and industry bodies carried out studies to determine the appropriate proportion of Robusta and Arabica to be cultivated in line with global demand. Accordingly, Robusta plantations were replaced by Arabica.The aim is to achieve a proportion of Arabica and Robusta as 75:25,to align it with global demand.Intensive farming and more advanced technologies have led to a consistent increase in paddy yields from 3.69 tons/ha in 1995 to 4.6 tons/ha in 2003; higher than other Asian countries including India (2 tons/ha),Thailand (2.2 tons/ha), Myanmar (3.2 tons/ha) and the Philippines (2.89 tons/ha). Inviting foreign investment in production and trading of coffee - many of the world's largest trading houses in the world such as ED and F Man, Newman Groupe and O Lam are present in Vietnam.

Promoted application of international standards to domestic coffee. Rice Increased competitiveness through higher productivity: High competitiveness in rice has been achieved on the back of consistent increase in area and productivity for paddy. The cashew processing industry in Vietnam has made significant contributions to enhance exports. The developments in processing technology have enabled Vietnam to export cashew in processed form. Vietnam has developed "Cover split technology-designed by Vietnamese technicians. This technology is cheap and is able to generate a higher ratio of whole seed.Due to easy availability of efficient technology, the number of processing companies increased from 6 in 1986 to 30 in 1994 (with total capacity of 75000 tons/year) and to 62 in 1999 (with total capacity of 250000 tons/year) to about 120 in 2003. These measures have facilitated the increase in output from 19.2 million tons in 1990 to 31.3 million tons in 1999, and further to 34.4 million tons in 2003.

TUNA FROM THAILAND MARKET AND PRODUCT DEVELOPMENT

Thailand is the world's largest producer, and second largest exporter of canned tuna. 90 per cent of Thai tuna production is exported, bulk of it having been imported from other countries in raw form. The export value of Thai canned Tuna doubled in the last ten years to about THB 30,000 million (USD 670 million).Canned tuna from Thailand faced severe competition in its key markets viz the USA and the EU in 2002.

Both the USA and the EU have allowed duty-free access to canned tuna from Andean countries. Further, the EU has also provided duty free access to the ACP (African, Caribbean and Pacific) countries. The duty rates imposed on tuna by the USA, the EU and Japan are tabulated below.

- Dependence on imported supply and fluctuations in world prices of tuna and tin.
- Diversify base of suppliers.
- Financial instruments to mitigate the risk of fluctuations.

Trade:

- Develop new export markets (China, India).
- High quality production processes and safety measures (for, *e.g.*, canneries shifted to sunflower oil after genetically modified soyabean oil was banned by some importing countries).
- Non-trade barriers environmental concerns, GMO issues, food labeling and safety.
- Trade barriers import quota, tariff.

Competition:

- Competition from ACP and Andean countries.
- Development of domestic market for canned tuna.
- Partnership with global tuna companies diversi-fication into other food businesses such as pet food, fish oil to safeguard against cyclically.

Substitute products:

- Product development, *e.g.*, tuna-based ready to eat, processed foods and snacks and other animal protein products.

- Threat from other seafood products.
- This strategy enabled Thailand to maintain its leadership position in tuna exports.

DEVELOPMENT OF DOMESTIC MARKETS

The key markets for Kenyan tea are Pakistan, UK, Egypt, Afghanistan and Russia. In the recent past, two important markets-the UK and Egypt, reduced imports of tea from Kenya. The Kenyan Tea board decided to broad base its markets and started promotion activities targeting new markets such as West Africa, Eastern Europe and the Middle East. The Board is involved in both domestic and interna-tional generic promotion of Kenya tea through participation in Fairs, Symposiums, Seminars, etc.

A sustained generic promotion campaign was launched by the Kenyan Tea Board in October 2002 to popularise tea-drinking in the country. The above efforts have contributed significantly to growth in demand in the local market tea sales grew by 14 per cent to 14.4 million MT in 2003 after declining for 10 years. The campaign, in its second year, spurred rene-wed brand promotions by various packers to ride on the increasing tea and health awareness created by the Board throughout the country. The Generic Tea Promotion Campa-ign aims at increasing the local per capita consumption from the current 500g to 805g in the next five years. This will reduce Kenya's dependence on overseas markets.

Direct Farmer Processor Linkages and Operational Efficiency

Brazil is an outstanding success story in agri and food exports, having achieved leadership status in a wide range of agri products including sugar, orange juice, meat, oilseeds, etc. Brazil's agricultural and food exports constitute 34 per cent of total exports from the country and are valued at USD 25 billion. The key underlying factors which have enabled Brazil to achieve the above are as follows:.

- *Direct farmer processor linkages*: This has had multi-fold benefits including enabling processors to achieve scale in operations which has led to development of sustainable business models, ensuring farmers adopt best practices to enhance crop productivity and availability of financing to growers, on the back of firm off take arrangements with processors. The other key enabler is the free market system which has ensured that processors and farmers aim to maximize operational efficiencies. For, *e.g.*, one ton of sugarcane produces 140 Kg of Sugar (vis a vis 100 kg in India,85 kg in Argentina) in Brazil which is among the highest in the world.

New Zealand - Brand-building in Kiwifruit

Zespri's success in marketing of kiwifruit is well established. The organisation has been able to achieve this through a consumer-led strategy, while adopting a holistic supply chain approach to address various bottlenecks and issues in the areas of research, distribution and marketing.

VISION, STRATEGY AND ACTION PLAN FOR AGRICULTRAL EXPORTS

- *Vision*: The reduction in subsidies by the EU and US will lead to the emergence of new opportunities for exports of Indian food products over the next ten years. Supply chain efficiencies together with a focused approach to enhance exports are key to ensuring that India is able to successfully tap new product/market opportunities. India has the potential to achieve a 3 per cent share in world trade of agricultural and food products by 2015.
- *Strategy*: In order to sustain the momentum of penetrating new markets without losing share in existing markets, it is important to focus on key markets, which offer high growth potential and products where India has an inherent production advantage.

PRODUCT MARKET FOCUS

As highlighted in the case studies, "focus" and "long-term strategy" are the cornerstones for sustainable success in agriculture and food exports. It is recommended that the following products/markets be focused on, in the initial phase.

These product categories have been identified on the basis of one or more of the following parameters:

- Comparative cost advantages.
- Current and likely trade volumes in the category, based on underlying demand trends.
- India's production advantage (in aggregate terms or for specific varieties).
- Potential for differentiation.

Product Segmentation

India's exports of food and agricultural products can be categorised on the basis of market potential and the oppo-rtunity to differentiate the product offering. The opportunity to differentiate emanates from inherent production adva-ntages on account of low cost production and/or availability of unique varieties. For instance, India has unique varieties such as Durum wheat, Darjeeling tea, Alphonso Mango which can be leveraged effectively to establish a significant presence in exports, both in terms of absolute value as well as market share. On the other hand, in dairy, India has significant cost advantages, which can render it a significant exporter, (provided quality challenges are addressed) especially in view of declining subsidies in the EU.

Improvement in Market Access

Two key steps for improving market access of Indian food products in overseas market is to institutionalise the market intelligence network for exporters and harmonisation with international standards and develop associated infrastructure for certification and testing

Market Intelligence

In order to focus on some key products and markets, it is critical to develop a strong database to enable current and potential exports to take rational decisions.

The key information needs of exporters include:

- A special cell needs to be set up by the Ministry of Food Processing which provides requisite info-rmation on the above aspects to exporters.
- Current status of quality standards and food regula-tions in target markets for imports of identified pro-ducts.
- Existing tariff structure and non- tariff barriers, and likely changes in the context of WTO requirements.
- India's competitiveness vis a vis key competitors.
- Major importing markets.
- Harmonisation with international standards/practices, certification and testing.

One of the major challenges for India, following the dismantling of quantitative restrictions on imports, is to raise the level of quality standards to become globally competitive. There are variation in standards and regulations adopted by different importing countries, which may lead to trade conflicts and disputes.

The specific steps in this direction are:

- Encourage food testing laboratories in India to obtain accreditation from international agencies. Given high cost of international accreditation, Government can incentivise laboratories by part funding these costs.
- Encourage importing countries (primarily USA, EU, Japan) to set up offices in India for certification of export consignments
- Expansion of the list of export products for certification by EIC (Currently, EIC certifies six notified commodities *viz.* Basmati rice, Black pepper, Egg products, Honey, Fish products and Milk products).
- Introduce certification zoning systems pesticide free zones, organic production zones, disease free zones to facilitate high value exports from India.
- Promote certification for organic farming for different crops.
- Substitute post arrival testing of Indian products in the importing country with pre-shipment inspection reports by recognised international agencies.

6

Globalization and Indian Agriculture

In my treatment of the vast theme of globalization, I will limit myself to the agricultural sector of my country, India. My thesis is simple. And although no one should believe a person who begins by claims to simplicity, I argue that until recently, globalization in Indian agriculture took the form of international cooperation. Further, this collaboration constructed the base for the recent entry of private, for-profit foreign investment. I also argue that this earlier globalization was indeed effective, together of course with steps taken within the country, in "cultivating peace" that is, in a country with many reasons for violence, agricultural prosperity indeed averted large-scale conflicts. In contrast to that period of internationality, the current globalization risks much if it does not include a strong, sustainable plan for those who are the most vulnerable in the processes of development.

The Earlier "Globalization". International collaboration in Indian agriculture had already begun in 1957 when the Rockefeller Foundation and the Indian Government began a programme to improve maize. Because of this early link with foreign collaborators, the country received varieties of wheat and rice from, respectively, Mexico (www.cimmyt.org) and the Philippines, (www.irri.org) in the early Sixties. Five years later in India, the Green Revolution was launched (which, alas, is not at all related to the Green of the Esperanto flag, I must quickly clarify! The technological package included better, higher-yielding crop varieties; use of chemical fertilizers; modern irrigation systems; and strengthened programmes which supplied farmers with practical knowledge.

Also at that time was established in the southern Indian city of Hyderabad the institute where I now work, ICRISAT (www.icrisat.org) — the English acronym for an internattional research institute for agriculture in the semi-arid tropics. This institute, with the 16 others which are members of the Consultative Group for International Agricultural Research have contributed much to agricultural growth not only in India but also in the rest of the developing world. For example, in the 25 years of its existence, ICRISAT's partnership with national agricultural systems developed around 370 varieties of grains and legumes for distribution in more than 70 countries; collected and conserved from 130 countries, the seeds of more than 113 *thousand* varieties of these six grains and legumes; and ICRISAT has trained more than 3000 scientists from 90 countries.

A Strong Infrastructure. This kind of successful collaboration presupposes the existence of a sufficiently strong infrastructure for agriculture, which India indeed possesses. Of course, in a country where about 65% of the population finds a livelihood in agriculture and related sectors, any general raising of standard of living must necessarily involve the rural sector.

The so-called "Five-Year Plans" of the economy (from 1951 onwards) have, in fact, done exactly that. Much investment both human and financial was made in agriculture. The great part of this investment was capital investment, for example, roads, and large irrigation projects. This large-scale construction of agriculture infrastructure too received much support from the "first world". For instance, in the 25 years between 1962 and 1987, just the irrigation sector received 3800 million dollars from the World Bank.

India also received help in the design and implementation of policies which aimed to protect the most vulnerable sections of the people. The main policies included: price-support for farmers; establishing "fair-price shops"; policies regarding the procurement of agricultural produce; controlling the flow of food across state boundaries; and a flexible import policy that allowed large imports when it became necessary. This combination of infrastructure, science, and policy indeed averted both catastrophic famines of the earlier kind, and large-scale unrest. And reasons for unrest there were (and indeed are!) aplenty: India is home to 15% of the world's population; and to a quarter of the population of developing countries. As already mentioned, 65% Indians live in the rural sector where poverty is widespread: 53% of the inhabitants have less than 1 dollar a day to spend.

Unjust Land Distribution. The rural violence which did occur during the last 50 years of Independent India of course has complex roots in history and culture, but that violence also shows how easy it is for violent conflict to occur in underdeveloped regions.

The "decade of chaos" from 1967 to 1977 in West Bengal was caused by a serious inequality in land distribution. Almost half of the population either did not possess any land, or had less than a quarter of a hectare. The landlords constituted less than 7% of the population.

Further, hordes of refugees entered the state during the war in East Pakistan. That caused acute food shortages in the state. The tragic violence of that period forced the communist government to change the policies of land distribution, and those changes brought many people out of the so-called "Naxalite movement" into the mainstream of civil society.

The process of just land distribution is far from done: less than 1% of the cultivable land in the country is fairly distributed. And this unjust land-distribution is only one of the many points in the so-called "unfinished agenda" of the developmental process. I will list other elements of that "unfinished agenda" later, but the point to emphasize here is that the current globalization offers no possibilities to address these social needs.

The Current Globalization. The current globalization started in 1991 during a financial crisis in the country. A new government stabilized the economy, simplified investment in the country, removed many trade barriers, and reformed the tax-system. On account of this liberalization of the economy, direct foreign investment in the country increased more than 10-fold between 1991 and 1996. Further, in the agricultural sector, this opening up of the economy combined with a series of good monsoons causing rapid agricultural growth from –2.3% in 1991 to 4.9% in 1995.

But this financial blessing evidently happened only for the relatively rich farmers. The fate of the poor did not improve. For rich farmers globalization means, for example, that they have access to new hybrid seeds. Since these hybrids cost much more than the traditional low-yielding varieties, only the rich agriculturists can buy them. Further, none of the private firms "waste" money in improving the so-called "orphan plants" — grains like sorghum, and legumes like chickpea. Globalization — the opening up of the market — does not mean improvement in the lives of marginal farmers.

The Dangers. There are also other, more general, dangers for agriculture in the current globalization. I will quickly touch on only two, namely, intellectual property rights, and biotechnological considerations.

Intellectual Property Rights in agriculture are supposed to protect both the rights of firms which improve crops, and the rights of agriculturists who possess crop-varieties with valuable traits. In practice, however, there is almost no awareness in rural India about the rights of farmers. Meanwhile, private firms patent crop varieties, as happened recently when researchers in Australia applied for a patent for varieties of chickpea.

They claimed that although they had obtained those varieties from the ICRISAT genebank, they improved the seeds and so have a right to patent them. Only the alarm sounded by a nongovernmental organization woke the institutions of the CGIAR and the Food and Agriculture Organization of the United Nations which immediately placed a moratorium on the patenting of these crops, defining them "international public goods".

Biotechnology firms present another aspect of the dangers of the current globalization. These firms invest much in the testing and introduction of new hybrids which both cost a great deal and whose ecological consequences are insufficiently known. Further, these firms protect their investment in ways which are ethically indefensible.

The American biotechnology firm, Monsanto, for example, recently found itself in the centre of a political storm on account of its attempt to introduce rice seeds containing the so-called "Terminator" gene; that is, that high-yielding rice hybrid contains a gene that makes sure that seeds from this season's rice crop will not germinate when they are sown the next season: the farmers therefore must again buy seeds from the firm.

Well, once again, a nongovernmental organization sounded the alarm, and a gravely embarrassed Monsanto hastily declared that the firm was only testing the possibilities of the Terminator gene, and had in fact no plans to use it. In both cases, then, nongovernmental organizations played an important role. On the one hand this emphasizes the role of a free media, and thus, a democratic society (a point which the economist Amartya Sen stresses). But on the other hand, this phenomenon also indicates a disturbing abdication by the state of its duty to protect the most vulnerable sections of the population.

The Heart of the Problem. And that takes us into the heart of the problem with globalization. The state continues to have an important role in a globalizing world — arguably an even more important role than in earlier times when, for example, India was economically impenetrable. But the current economic climate does not encourage one to hope that this realization will be generally accepted. Further, the political instability of coalition governments, and extremely costly "distractions" such as for instance the recent "almost-war" with the neighbor Pakistan, make the task of the state in the developmental process even more difficult. Meanwhile, one out of three Indians does not have access to safe water; almost one out of two illiterates in the world is an Indian; the largest number of the absolutely poor live in India; the country has more registered unemployed than the total unemployed of the 24 countries of the OECD; and 44 million children in India work.

The list of this "unfinished agenda" continues. Meanwhile, as the 1999 Human Development Report (www.undp.org/hdro/report.html) of the UNDP says: "when research priorities are defined, money, not need, rules — cosmetic drugs and slow-maturing tomatoes appear higher in the priority list than drought-resistant crops and vaccines against malaria."

Only if a democratic government both fulfills its role and forces private firms to participate actively and positively in the process of social development, will globalization not have catastrophic, violent consequences. Only then will be realized the UNDP's aim of "Globalization with a Human Face."

THE ROLE OF THE STATE IN AGRARIAN TRANSFORMATION

In India, the role of the state in bringing agricultural transformation is made possible by active co-operation between millions of farmers across the country and the governments at the federal and provincial levels. As we observe elsewhere in the text, agriculture in India is policy driven. According to Singh (1995) "the broad policy framework towards agricultural sector in India consisted of the following:

- Land reforms to remove the intermediary landed interests coupled with measures like consideration of holdings and some half-hearted programmes of land redistribution.
- The creation of supporting infrastructure in the forms of irrigation facilities, rural roads, electrification, credit and market facilities etc.
- Creation of an agricultural research, training and extension system on a vast scale, and
- A system of public procurement and distribution of agricultural products, particularly food grains with a judicious balance between the interest of the producers and consumers".

It is agreed amongst many professional economists that agriculture should never be left entirely at the mercy of the free market.

State intervention in agricultural markets can be justified on various grounds:

- Unlike industry, where production is a continuous process, agricultural output appears not continuously but at discrete intervals, because of which output cannot be adjusted to demand conditions with any rapidity.
- The ability of peasants to hold back their produce/ stocks is limited in the case of agriculture as the scale of operations is not as big as in industry.
- There is large fluctuation not only in the supply adjustments, but also in the overall output induced by weather and other natural calamities.
- Most importantly, demand for agricultural produce is price-inelastic, which means that agricultural prices and incomes are subject to large fluctuations.

Thus, the responsibility of the state is to ensure that adequate public investments are made in irrigation, agricultural research and infrastructure that ensures well-functioning markets and would facilitate exports. Parikh *et al.*, (1999) point out that "for food security, the government should follow a mixed strategy of modest buffer stocks supplemented by occasional trade on the world market.

We must note that agriculture is basically a private activity in which public investment has a critical role to play in providing the infrastructure such as roads, electricity, technology, storage facilities, etc. This not only requires high investments on part of the government but also requires proper direction and focus.

Thamarajakshi (1999) notes that "in this context GCF (Gross Capital Formation) in the public sector in agriculture, both in absolute and relative terms has been rather low whether in the pre-reform or reform period besides recording a decline in the later period; real GCF in agriculture declined by 14.3%

between 1985-97 compared to an increase of 13.3% in total public sector GCF while GCF in private sector rose in agriculture, the rise was lower (58%) than that in non-agriculture (74%) apart from the fact that the absolute size of private sector GCF in agriculture, although much higher than public sector GCF in agriculture, was comparably quite low in relation to the size of private sector GCF in non-agriculture".

Table. Gross Capital Formation in Public and Private Sectors

(Rs. in crores)

	Agriculture	*Non-Agriculture*	*Total*
Public Sector			
1985-91	11347	18229	19576
(6.9)	(93.1)	(100.00)	
1991-97	1155	21020	22175
(5.2)	(94.8)	(100.00)	
Private Sector			
1985-91	2995	19891	22886
	(13.1)	(86.9)	(100.00
1991-97	4736	35195	39931
	(11.9)	(88.1)	(100.00)

Note:(i) Average for each period at constant, (1980-81) prices.;
(ii) Figures in parentheses are percentage total.
Source: Thamarajakshi, 1999:2293.

Thus, in India with the state's emphasis on a market-based system under liberalisation, the overall public investment relating to total investment in agriculture has been constantly declining. As is evident from Table, there has been a decline in GCF in agriculture vis-a-vis non-agriculture. Also there was a decline both in public as well as private investments. An interesting point to be noted is that private investments in agriculture are more than double than the investments by the public sector. However, during 1993-98, corresponding GCF for public and private sectors was around 6% and 9% respectively. Decline in private sector GCF in agriculture under liberalisation is conspicuous. According to a report by the Planning Commission (1997), planned investment during 1992-97 (*i.e.*, precisely after completion of Eighth Five-Year Plan) in agriculture was over Rs. 6,49,999 crores, while actual investment was only at Rs. 3,83,000 crores (at 1996-97 prices). The distinction between actual investment to that of planned Public investment in agriculture being 59%(5) (Planning Commission, 1997:12).

Table. GCF in Agriculture (% to total GCF at 1993-94 Prices)

Year	*Public Sector*	*Private Sector*	*Total*
1993-94	6.3	10.7	8.9
1994-95	6.1	10.1	8.9
1995-96	6.4	89.0	8.6
1996-97	6.0	9.4	8.1
1997-98	5.6	8.9	8.3

Note: Percentage in each column is that to total GCF in the respective sectors.

We know that in many areas of India, irrigation forms the basis for agricultural growth. Investment in this sector is crucial for acceleration of growth in agriculture. The total investment in irrigation under liberalisation was around Rs. 5,15,000 crores, which was slightly higher (around 4%) than the earlier period. According to a recent issue of Economic Survey (1999), area under medium and major irrigation increased by 2.1 million hectares during reform period, in comparison to near constancy earlier. However, minor irrigation recorded highest increase of around 14.1 million hectares for the period 1985-97.

Table. Investment in Irrigation (At constant 1996-97 prices)

Power	*Major & Medium*	*Minor*	*CAD(*)*	*Total*
1985-92	29332.8	1630.2	3684.9	49325.9
1992-97	31057.6	17302.5	3162.9	51523.0

(*) Catchment Area Development.

Source: Planning Commission, Nineth Five-Year Plan.

FACTORS CONTRIBUTION TO DECLINE OF AGRICULTURE

Slow Down in Agricultural and Rural Non-Farm Growth: Both the poorest as well as the more prosperous 'Green Revolution' states of Punjab, Haryana, Andhra Pradesh and Uttar Pradesh have recently witnessed a slow-down in agricultural growth and it ultimately lead for farmer's suicide. Some of the factors hampering the revival of growth are:

- *Poor composition of public expenditures*: Public spending on agricultural subsidies is crowding out productivity-enhancing investments such as agricultural research and extension, as well as investments in rural infrastructure, and the health and education of the rural people. In 1999/2000, agricultural subsidies amounted to 3 percent of GDP and were over 7 times the public investments in the sector.
- *Over-regulation of domestic agricultural trade*: While economic and trade reforms in the 1990s helped to improve the incentive framework, over-regulation of domestic trade has increased costs, price risks and uncertainty, undermining the sector's competitiveness.
- *Government interventions in labour, land, and credit markets*: More rapid growth of the rural non-farm sector is constrained by government interventions in factor markets — labour, land, and credit — and in output markets, such as the small-scale reservation of enterprises.
- *Inadequate infrastructure and services in rural areas*. Infrastructure is also a significant factor in the process of development but country like our rural Bharat has not posses the infrastructure such as roads, electricity, fertilizer and pesticides availability which caused the vulnerable damage to the growth of agriculture.

Weak Framework for Sustainable Water Management and Irrigation:

- *Inequitable allocation of water*: Many states lack the incentives, policy, regulatory, and institutional framework for the efficient, sustainable, and equitable allocation of water.

Deteriorating irrigation infrastructure:

- Public spending in irrigation is spread over many uncompleted projects. In addition, existing infrastructure has rapidly deteriorated as operations and maintenance is given lower priority.

Inadequate Access to Land and Finance: Stringent land regulations discourage rural investments: While land distribution has become less skewed, land policy and regulations to increase security of tenure (including restrictions or bans on renting land or converting it to other uses) have had the unintended effect of reducing access by the landless and discouraging rural investments.

Computerization of land records has brought to light institutional weaknesses: State government initiatives to computerize land records have reduced transaction costs and increased transparency, but also brought to light institutional weaknesses. Rural poor have little access to credit: While India has a wide network of rural finance institutions, many of the rural poor remain excluded, due to inefficiencies in the formal finance institutions, the weak regulatory framework, high transaction costs, and risks associated with lending to agriculture. Weak Natural Resources Management: One quarter of India's population depends on forests for at least part of their livelihoods.

A purely conservation approach to forests is ineffective: Experience in India shows that a purely conservation approach to natural resources management does not work effectively and does little to reduce poverty. Weak resource rights for forest communities: The forest sector is also faced with weak resource rights and economic incentives for communities, an inefficient legal framework and participatory management, and poor access to markets.

Weak Delivery of Basic Services in Rural Areas: Low bureaucratic accountability and inefficient use of public funds: Despite large expenditures in rural development, a highly centralized bureaucracy with low accountability and inefficient use of public funds limit their impact on poverty. In 1992, India amended its Constitution to create three tiers of democratically elected rural local governments bringing governance down to the villages. However, the transfer of authority, funds, and functionaries to these local bodies is progressing slowly, in part due to political vested interests. The poor are not empowered to contribute to shaping public programmes or to hold local governments accountable.

AGRICULTURE EXTENSION IN DRY-LAND

The agricultural management systems in India and demonstrated the importance of considering multi-strategy solutions in agriculture development.

The severe environmental constraints in dry land agriculture merit a dynamic approach to agriculture improvement that encompasses not only nutrient management and agro-forestry, but particular attention to local watersheds. Though governmental extension services in India have adopted agroforestry and some watershed management programmes in their extension training, little emphasis has been placed on human and environmental development where decisions and action plans are decided at the community level.

Public extension in India does not have the resources to implement integrated programmes that value the participants' development even more than the farm systems' development. In response, private extension, primarily in the form of non-governmental organizations (NGOs), has begun to expand beyond the reaches of the production-oriented public sector, both in regional coverage and methodology. Integrated conservation and population programmes are based on the principle that human health and development is essential to the sustainability and the protection of ecosystems, natural resources, and communities. An analysis of the global relevance, the principles of human and integrated development, their advantages, and a case study of community based population and environment programme elucidates the magnitude of change integrated programmes can generate.

By nature, communities are the interface between humans and the environment. At this interface there are numerous factors, including population, socio-economic, political, biophysical, and cultural components that influence the lifestyles of community members and their interactions with the environment. Historically, India's policies, aid programmes, and extension have dealt with these development issues through separate ministries, rather than approaching them from the viewpoint that people and the environment are interwoven.

Globally, the existence of this linkage between human health and the environment was acknowledged at the United Nations Conference on Environment and Development in 1992. The primary health care and environment millennium goals determined at this conference clearly state the international community's acceptance of this linkage between the environment and human health. Since then, efforts have been made to discover, design, and/or analyze initiatives and programmes that combine community-based population and environment.

Though the global community and perhaps the Indian government have acknowledged the logic of combining health and environmental services, both are far from implementing such programmes. Limited public resources coupled with burgeoning problems and dogmatic development strategies prevent more adoption of integrated community development programmes in India. The private sector has started to fill this neglected area, focusing on human development as an essential step towards sustainable and scaleable programmes. The inherently interwoven framework of communities and human interactions with the environment attest that agriculture development is limited without human development. Though agriculture development may improve the farmer's production, increase his income, and preserve the land for future generations, the sustainability of these results and the expansion to other individuals and communities may be compromised if health and socio-economic conditions are ignored and the people themselves are not taught self-reliance.

There are several conceptualizations and methods of achieving human development that vary between organizations, but most share root convictions and the final goal is always self-reliant, confident individuals and communities capable of making informed choices and decisions. According to Don Jose Elias Sanchez, whose work with poor farmers in Honduras is documented in the *The Human Farm,* human development grows from the conviction that only strong hearts and heads can guide the hands. Sanchez's training farm is centered around exercises and sessions designed to help farmers realize the strengths and understanding they bring to agriculture. This ground-up, holistic approach develops human capacity in a working context that farmers can transfer back to their own communities. UNICEF on the other hand, perceives human development in three basic principles whose achievement varies across place. The first principle is "sustainable human development evolves from a self-reliant understanding of local needs and resources."

Communities have an intuitive sense of the dynamics of the local ecosystem, economic potentials, market forces, and social development variables. Recognizing these dynamics and giving community members the skills to collect data, respond to problems, and analyze situations improves self-esteem and can lead to a more sustainable project. The second human development principle is "action must grow from a combination of a bottom-up and top-down programming." The informed decision-making needs to be based on a three-way inclusive dialogue between the *entire* community, government and agency officials, and experts. The third principle is "sustainability is only possible when action grows from community participation and self-reliance."

The success of an agriculture development project is contingent on the community's ability to analyze and respond to new problems after the withdrawal of the development agency. If communities are not taught the skills to understand and evaluate their own environment, the likelihood of expansion and/or sustainability is diminished. Here, in the third principle, is where the ideas of *The Human Farm*, UNICEF, and most human development schemes converge: the development of humans themselves empowers the individual and community, enabling the expansion and growth of their potential through their self-reliance and independence. This form of development is key to the future success of extension in India; the investment of time and training in people would assist the government in addressing the multitude of problems rural Indians encounter. The case study presented later in the paper elaborates on human development in the Indian context.

The integration of health and environment is advantageous to community-based extension because it addresses community interests as a whole, rather than as separate problems that are exclusively independent. Two of the foremost issues that need to be addressed in rural India are public health and agriculture. The leading causes of death in India include neo-natal, diarrhea, and pneumonia; infections and diseases that are all exacerbated by malnutrition. Anemia, a micro-deficiency caused by lack of iron, is prevalent in more than 69% of pregnant women in India (CDC, 2000). Vegetarian diets, lack of protein and bio-available iron sources contribute to a deficiency that has severe consequences such as poor brain development in children, compromised immunity, maternal mortality, and reduced work capacity.

The majority of these deaths and illnesses could be prevented with proper health interventions and food supply improvements. The health of people is not contingent only on the availability of food, but the sanitary conditions and practices of families. Primary health care such as access to rural heath posts, education on proper infant and childcare, maternity care, reproductive care, portable water, and sanitation are all basic needs and rights the greater part of rural Indian people are denied. Beyond the logical sentiment that this is iniquitous, researchers at the forefront of nutrition and economics argue child and adult malnutrition reduces labour output, GDP, cognitive ability, and motivation. It is not only beneficial to improve the health standards of people for their own wellbeing and ethics, but for the wellbeing of the country.

Agriculture and public health are linked in two ways. First, without sustainable farming techniques, diversified crops that bring in multiple nutrients, and income for supplemental foods, the consistent yield, sustained nutrition and capital of the farmer and his or her family are compromised. Second, poor health, whether in the form of micronutrient deficiencies or parasites, can contribute to an inability to address farm issues with the energy and vigour expected from a healthy farmer and family.

This linkage of health and motivation is not only an example of the significance of linking health and agriculture, but it further validates the need for health in human development programmes. Linking health and agriculture has the advantage of addressing these principal issues at the same time. Furthermore, while it may be incongruent for experts or professionals to tackle health and agriculture together because they cross disciplines, it is natural for rural communities to perceive both as interconnected and integral to their personal growth. Communities are affected by multiple factors, and a holistic perspective is intuitive to them. Advantages of holistic approaches to community development reach further beyond the more evident reasons discussed. This health and environment approach to community development can facilitate entry into communities. Projects that linked family planning and environment often found barriers were more easily overcome when environmental services provided the means of introducing family planning to the community. Reproductive health is a potentially sensitive topic in India. These programmes can generate confidence and community acceptance, in addition to being a way of creatively linking reproductive health to agriculture thus reducing potential hazards of discussing family planning. Environmental metaphors may include comparisons between the importance of spacing crops and the spacing of children or the need to prepare for children just as the land needs to be prepared for seeds. Additionally, the link has provided a gateway for women into the environment sector and men into the reproductive health arena.

Community-based population and environment programmes can meet the needs of remote communities more efficiently than most sectoral, public sector, and advanced technology driven programmes. The people that need health and environmental services the most are those living in remote areas with little access to roads, schools or health services. These areas are usually the last to receive government services, the least likely to accept such services when they are available, and the slowest to change attitudes and behaviors. As a result, these areas in particular require strong community based holistic approaches to meet varied needs and improve adoption through the establishment of trust and the thorough understanding of community culture. Integrated and human development programmes have the potential to actually generate human capital in these resource-limited regions.

This development of people is exceptionally important, as it is unlikely that further assistance will be readily accessible on the completion and withdrawal of the project. The strong base of combined health and environment programmes gives the organizations the latitude necessary to establish this trust, understand the communities, and improve self-reliance in these remote areas. There are other advantages to integrated development programmes, including the inclusion of men and women in typically gender sensitive topics, cost effectiveness, and sustainability. The case study will demonstrate these advantages and those already discussed. The work of World Neighbors, an NGO presently working in India that has approached development holistically with an emphasis on health, environment, and specifically human development.

World Neighbors (WN) works in Karnataka, a dry land state in southern India. Though the state is considered at the median level of development in major sectors, the Human Development Report 1999 "describes the regions covered by WN programmes as 'areas of darkness' with HDI indicators that fall below the sub Saharan African countries (WN, 2005)." The region is typical of many dry land areas around the country, with agriculture depending on scanty and erratic rains and over 90% lacking irrigation (WN, 2005). As a result, farmers and shepherds of the region often migrate in search of food, fodder, and employment (WN, 2005). Malnutrition and infant mortality rates are high in comparison to other areas in the state (WN, 2005). In 1998, WN began working in reproductive health in two villages; Nellur and Yelenavadgi, where agricultural activities focused on dry land agriculture, sheep health, and wool processing were already ongoing. Over the past three years WN has partnered with 8 additional NGOs to facilitate activities and trainings in order to meet several objectives:

1. To organize target groups to undertake development activities
2. To create employment opportunities for the rural poor and establish training centres for self employment for youth and women
3. To plan and implement various rural employment and development programmes with the cooperation of local groups for the all round development of the people
4. To sustain all development activities targeted to the groups (McKraig, 2002).

As the objectives demonstrate, WN work in Karnataka focuses on human development in order to create sustainable, fruitful ventures, regardless if it's in the agriculture, health, or micro-credit sectors. One particular aspect of the programme targets the formation of women's groups through a three-step process. The first phase involves WN and the NGOs holding initial discussions with agriculture groups about the formation of women's groups in the villages. Groups are established and reproductive health workers (RH) are identified and trained. In the second phase, the capacity building of women's groups with the help of the RH worker takes place.

The initial activities of the groups include the selection of leaders and training of leaders in such topics as group management, decision-making, recording keeping and monitoring and evaluation techniques.

In the third phase, women's groups conduct activities independently, the RH worker visiting the groups only every three or six months. The first two phases involve the intensive support of WN and the partner NGO to help establish the women's groups, develop capacity, and implement activities.

These steps are essential to the successful development and coordination of the groups. Only when the skills and capacities of the groups are firmly established, about a three-year process, do WN and the NGO withdraw further and the women's groups conduct the activities independently. Though the phases may be lengthy and intense, evaluation of the programme demonstrates that women are empowered in many aspects of their lives.

An evaluation of the reproductive health component of WN's work was conducted in 2001, with indicators suggesting the human development methodologies improved the capacity of women to make informed decisions and the integration of health and agriculture was beneficial. Key findings from the evaluation state that "significant changes were reported in key reproductive health practices and rates for service use," compared to non-intervention villages. Additionally, "there were reported changes in indicators of women's status (McKraig, 2002)." The women indicated that group activities improved decision-making, property ownership, girl child education, and reduction of violence against women." Women felt they were more confident "to organize and undertake a variety of initiatives ranging from closing bars and gambling houses to using civil disobedience to improve water supplies."

The linkage between men's agriculture groups and women's groups greatly contributed to the success of the women's groups. The agriculture group members assisted in the formation of the women's groups, encouraged their wives to attend, convinced other men to do the same, and supported them in terms of organization of events and discussions. The linkage also encouraged communication between the men and women about dry land agriculture and "improved awareness of women's group members regarding key agricultural practices." Another advantage of the programme is the integrated model appeared cost-effective. Not only did the programme improve the group member's livelihoods, but spread to non-group members. Finally, the inclusion of trainings in evaluation, monitoring, and implementation techniques helped ensure the sustainability of the programme. "The women's groups are confident in their ability to sustain key activities after the phase out of BSRDS." The development of human ability encourages the scaling up of a programme; it is a positive step towards the expansion of the programme to other communities.

The evaluation of the reproductive component of WN's work in Karnataka demonstrates the value of integrated programmes. WN states on their Karnataka webpage, "the most important determining factor of success is the involvement and integration of various community groups (WN, 2005)." The apparent exclusivity of health and agriculture to some professionals is in direct contrast to rural community's instinctive perception that health, agriculture, and a myriad of other issues are innately associated. Furthermore, the case study demonstrates that the development of human capacity enables communities to determine the focus of their projects and their own future. The development of human capital is one of the most efficient methods of any form of development as it empowers people to actually *do* the development themselves and enables communities and individuals to grow and expand beyond the original confines of their poverty. This form of development is part of a viable solution to the multiplicity of problems India faces.

AGROFORESTRY EXTENSION IN INDIA

Agroforestry has the potential to significantly improve the livelihoods, economic viability, and agricultural production of small farmers in traditionally marginalized areas. In addition, agroforestry can be used as a tool for changing attitudes towards natural resource management. In order to understand how extension can better accommodate natural resource management through the successful introduction of agroforestry and tree planting, extension agents, both public and private, must strive to understand the socio-economics of agroforestry and of the communities within which they work ("On farm...").

While agroforestry has traditionally been successful in humid and semi-humid areas, it is most needed in dry-land areas, which are characterized by acute climactic challenges and extremely low levels of agricultural production. The limits to the adoption of agroforestry are many, but must be viewed beyond technical and biotic constraints of the dry-land agricultural regions; the ability of extension agents to recognize socio-economic characteristics of farmers is essential in learning how to target potential innovators who can then generate change within their communities.

More emphasis on the development of participatory programmes, more efficient targeting, and intensified focus on the traditional communication channels will increase the adoptability of agroforestry in dry-land communities. In turn, an implicit understanding that the attitudes of farmers towards natural resource management are based on agroforestry successes and failures will help extension to increase the environmental reclamation of degraded landscapes, especially those in dry, rain-fed agro-ecosystems.

Dry-land agricultural areas in India face specific problems that must be overcome for both improved agricultural production and environmental conservation. Agroforestry can help overcome limited water availability, fuel shortages, degraded soils, and a lack of biodiversity ("Agroforestry Systems in India" 2004 and CRIDA). When trees are used that are adapted to water-limited areas, such as *Gliricidia*, communities can realize both agricultural and environmental benefits (CRIDA). In general agroforestry produces a multitude of benefits, including increased income from selling surplus goods, improved environments and enhanced biodiversity, increased access to education through income generation, shelter from harvesting trees for building materials, improved health when medicinal plants are incorporated, a greater diversity of food sources, and an increase in the energy of an agricultural system because manure can be used for fuel instead of food.

Specifically in rain fed agro-ecosystems, like those found in dry-land agricultural areas of southern India, successful agroforestry projects can increase the organic matter in soil, improve the physical properties, increase crop yields, increase the availability of nitrogen in the soil, increase the water-holding capacity of soil, and reduce soil erosion. In addition, there is a reduction in pollution because of a decrease in chemical fertilizers. Most environmentally significant, the biodiversity of an agricultural area is enhanced (CRIDA). Agroforestry's scientific benefits as listed above have been well documented within the humid and semi-humid agricultural regions, and the challenge for extension agents is to successfully spread these benefits in dry-land areas where, traditionally, farmers have been marginalized due to resource limitations (such as soil-fertility and water availability).

These benefits are of utmost importance in creating a sustainable agro-ecosystem in dry-land agricultural areas, and they cannot be realized without appropriate extension and education. Most importantly such extension and education must be localized and participatory in nature. In order for training and education to be localized, extension agents must understand the important role of socio-economic factors as well as the technical aspects of agroforestry, and therefore they must begin to focus more on the social and communication constraints of the specific villages and communities within which they work. Indeed the foundation of success is the "participation of the people for whom the forestry programme is intended".

This participation of local residents will help transform the extension programmes from the current "top down" structure, pioneered in "transfer of technology" methodology, to "bottom up" participant defined priorities and goals. In this way, participants are much more responsive to the education and training offered by extension agents. In fact, the potential participatory nature of agroforestry programmes in rural communities can be a useful vector for knowledge generation that incorporates local information and global science. Villagers and communities, depending on a number of variables, are continually gathering knowledge from various sources, and extension agents should do less to emphasize the strict dichotomy of global science and indigenous knowledge as it can be harmful to the adoption rates of tree management and agroforestry.

Such a dichotomy will ultimately only deepen the impression of 'otherness' of indigenous and rural people's knowledge and thereby reinforce the developed/ underdeveloped polarization that many of these same researchers and development experts are trying to overcome...Rather than solely emphasizing differences, scholars might accomplish more by seeking out points of congruence and interface between local realities and larger-level systems.

Ultimately, Brodt and others advocate that synergy instead of trade-offs be emphasized when extension agents use a variety of knowledge sources to advocate certain practices. Communities within dry-land regions of India have already developed knowledge concerning tree management; however, these knowledge systems are "open", thereby allowing an exchange of information which can be more conducive to innovation than enhancing a "closed system" through strict juxtaposition of local knowledge with global science. By allowing the participants to innovate beyond their previous knowledge capacities, extension agents are facilitating local ownership of knowledge, which enhances the success of programmes and promotes long term commitment on the part of the farmer and community.

Participant defined priorities and goals and an incorporation of local knowledge by extension agents must also be complimented by designing programmes that better target potential users, early adopters, and innovators. Programmes must be designed to address different family units, income levels, ages, land tenancy, farming experience, castes, occupations, urban contact, degrees of social participation, farm modernity, household assets, access to capital, and varying off farm incomes. Methods have been developed that, when using these and other variables, can successfully identify those farmers that will be "tree planters" within a subsistence farming community. It is essential that, before an extension programme is implemented, "the beneficial characteristics that might promote or impede local involvement in forestry programmes" are defined. Mahapatra and Mitchell have identified several key factors that make a farmer into a potential "tree planter" including "progressive attitude", "membership in village organizations", "wealth status", and "perceived risk concerning agricultural production." In addition to such characteristics, they also note that off farm income generation is generally very complimentary to the low labour inputs of agroforestry (or home gardens).

Most importantly, it has been shown that social or economic factors cannot account for the adoptability of tree planting and agroforestry; these factors must be combined in order to successfully target farmers who will plant trees. Another important note is that land area and tenancy are important, but mistakenly overemphasized when targeting farmers: "Classification of farmers by size of landholding, as is usually done by agricultural extension workers in India, is not suitable for farm forestry extension". This is especially important for agroforestry extension in dry-land areas where farmers are operating on very small plots of land—further supporting the spread of agroforestry beyond humid and semi-humid areas.

An exchange of information and knowledge generation are both critical in innovation and the success of agroforestry, and the model developed by Mahapatra and Mitchell also emphasizes this critical point by observing that membership in village organizations and contact with urban populations are essential indicators in targeting adopters and innovators. By better understanding how to target specific farm types, farmers, and communities extension programmes can shift away from the "conventional approach" of delivering standardized tree seedlings and vague management advice to all farmers. In this way they can concentrate their energy, time, and resources on specific farm types and farmers as well as identify the "uncertain" group that could either adopt or dismiss agroforestry. By helping change the attitudes of "fence riders" and targeting the farmers that are very willing to innovate, agroforestry programmes will see much more long term success and increase the adoptability of tree planting and, thus, natural resource conservation.

The participatory model of extension and education advocated throughout this chapter as well as the emphasis on targeting potential adopters and innovators is underscored by a better understanding of local knowledge dissemination and technology transfer among rural farmers in India's dry –land regions. When extension agents invest time in learning how villages and farmers communicate, they also see that participatory programmes and targeting initiatives are critical in establishing successful agroforestry farms; "the decision to adopt agroforestry was found to be determined by the farmers' attitude to agroforestry, which in turn was shaped by information received through farmer-to-farmer and farmer-to-extension contact. The mode of communication was important and, to be effective, needs to be customized for each target group". Glendinning has found that the most important channels of communication for farmers, when innovating and learning, are extension agent, neighboring farmers, and group meetings.

These channels of communication indicate that participatory programmes and better targeting by extension agents are appropriate ways to disseminate knowledge and change attitudes. Agents can target potential innovators who are capable of bearing more risk than other farmers. When these innovators generate both economic and environmental benefits through agroforestry, it is more easily communicated to neighboring farmers and members of group organizations. Neighboring farmers are more likely to adopt agroforestry if they see it has been successfully done by members of their own community. Successful demonstrations by villagers and other farmers decrease the perceived economic risk of agroforestry. "In regions where smallholders predominate [like dry-land areas in India], promoting change within the existing farming system may therefore require substantial diversion of resources in order to develop effective communication systems, skills, and media".

It is therefore equally important for extension agents to assume the role of facilitators when first entering communities; they can create or enhance pre-existing village meetings and farmer-to-farmer contact. Extension agents can take advantage of the local social structures to increase the "adoptability" of agroforestry. In addition for small landholders personal communication with neighbors and extension agents are the two modes of communication and knowledge sharing that are most strongly linked to potential adoption and tree planting. Understanding the communication channels and constraints will enable extension agents to enhance village communication and capacity building of farmers, which will, in turn, help demonstrate the low economic risk of agro forestry as well as the environmental benefits and improvements in agricultural yields.

Local knowledge, participation, and better targeting by extension agents are critical in establishing a long term commitment to agroforestry in dry-land regions of India. The importance of successful agroforestry extends beyond satisfying agricultural production and economic goals, it is also critical in establishing an understanding of why natural resource management and environmental conservation are important in the sustained productivity of ecosystems. The future of natural resource management depends on the attitudes of farmers which are distinctly shaped by the information they receive from communication channels guided by extension services. Extension agents in India have established the scientific base and technical background to support agroforestry adoption; however they have yet to master the socio-economic tools to analyze communities and enhance local social structures that enable the adoption of agroforestry and facilitate knowledge exchange and generation, critical components of successful agro-ecosystems.

If extension agents can use socio-economic analysis to better understand the local conditions in which they operate, the adoption rates of agroforestry will increase, and the overall environmental benefits will be realized. These environmental benefits are critical to sustaining agricultural production to feed a burgeoning population. One of these, nutrient management in soil fertility, is of overwhelming importance for cropping systems. Further examination of nutrient management in Indian dry-land agriculture is needed to understand the future of agricultural productivity, and the overwhelming importance of agroforestry in degraded landscapes. Such novel approaches aimed at better understanding local conditions and farmer capacities and attitudes towards adopting improved land management are also a useful tool for analyzing nutrient management issues in much of India's degraded dry-land agricultural soils.

IMPACT OF GLOBALISATION ON THE AGRICULTURAL SECTORS

Agriculture is the engine of most African economies and in recent years governments have become convinced that, by liberalising their economies, agriculture would prosper and provide the necessary growth to provide investment to improve the country's infrastructure, to form the foundation for industrialisation and to improve public services.

In the 1980s some African countries began to reform their economic policies. Internal conflict, however, has delayed reform in many ECA countries and they have only recently begun to liberalise their internal economies. Over the last twenty years there has also been an accelerating trend to liberalise trade on a global scale. The success of these global reforms and internal liberalisation measures (for those countries that have adopted them) has been patchy. Most African countries have found it difficult to compete with more efficient foreign agricultural producers and are suffering surges in imported products which compete with domestic production.

At the same time, the expected improvement in exports of these products has not materialised. This may be due, in part, to the difficulties of complying with the high quality standards required by many importing countries. The international market prices of almost all agricultural commodities have fallen to their lowest levels (in real terms) in living memory. This is due to overproduction encouraged by the export-orientated economic policies of competing producing countries.

The cut in agricultural subsidies in the developed world has reduced surplus stocks of food which has had the effect of reducing supplies available for food aid. The lowering of tariff barriers by consuming countries has offered more opportunities to exporters in developing countries. The exposure of agricultural production to foreign competition has forced some producers to become more efficient.

The dismantling of government-controlled marketing boards has stimulated the evolution of the private sector trading networks needed in a modern economy. Some actors in the agricultural sector have seen little benefit from the liberalisation process, however. Communities of small-scale, isolated farmers (which make up the majority of the population in many ECA countries) find it more difficult to obtain inputs and credit. Extension services have been significantly reduced and the value of their surplus production has fallen.

They are especially vulnerable to changes in production systems. The trend towards larger farms and plantations in the name of efficiency has marginalised many rural groups thus adding to the problem of unemployment, urbanisation and cultural disintegration. The effort required to address these problems and challenges will have to be made by many agencies. Trade agreement negotiators, government agencies, agricultural development and agricultural research organisations, NGOs and private-sector farmers', traders' and processors associations all need to be fully aware of the changing relationships between markets and all the different actors in agriculture, and shape their policies and programmes accordingly.

This report examines the main benefits and disadva-ntages of the trend towards globalisation and the libera-lisation of agricultural markets. It also attempts to offer some suggestions on how ECA countries might maximise the opportunities offered by a more open global trading system and how they might respond to some of the negative aspects of liberalisation. In particular, the report will highlight key issues that need to be addressed by all agencies involved in agricultural development in an increasingly globalised economic environment.

EARLY TRADE AND COLONIALISM

East Africa has a tradition of trade, especially with the Arabian peninsular and Southern Asia, going back several thousand years. Arab merchants developed trading links with many African kingdoms and established settlements on the coast before the 8th century AD.Trade routes were established into the interior of the continent to transport commodities such as ivory, gold, furs, gums and slaves, to the coast and Asian food products such as bananas and coconuts as well as some spices were introduced. European colonisation began in the 15th Century. The Portuguese charted the coastline and developed natural harbours for use in their trade with the East and they too began trading in African goods. The Portuguese were followed by the French and British and later by the Germans and Italians.Each colonial power began to explore further into the interior and to subdue the indigenous populations. The British, especially, recognised the potential for agriculture in the region and established plantations and farms. Large areas were colonised and territorial borders delineated. Cotton was produced as well as food crops and livestock. Local African agriculture was transformed with the introduction of crops from other parts of the world, notably those originating in the Americas such as sweet potatoes, potatoes, maize, cassava, peppers, tomatoes, papaya, jackfruit, cocoa, passion fruit, pineapple, sisal, cashew nuts, sunflower, groundnuts and tobacco. In the late 19th century the first railways were driven deep into central Africa.

In certain instances workers were transported to Africa, especially from the Indian sub-continent, to work in the plantations and as labourers to build roads and railways. Indians with an entrepreneurial background were able to establish businesses that grew to represent a significant proportion of trading activity in some countries. Europe's main commercial interest in ECA was its raw materials – minerals and agricultural products. The present transport system reflects those interests. Railways and major roads were designed to carry products from the interior to the coast, not to encourage or facilitate trade within the region.

Borders between countries are largely arbitrary, from an African point of view, and they often cut across the territory of populations with a common language and culture. European companies dominated import/export trade in African products often holding a monopoly in specific commodities. Many of these companies not only organised the production of these commodities but also transported them and processed them locally or, more often, in their home country. Some, like the giant chocolate, tea, sugar, tobacco and coffee companies, also marketed the finished product.In order to protect these monopolies trade barriers had to be erected between African countries colonised by different colonial powers. At times, countries of the region were officially at war with each other as the French fought the English, the English fought the Germans, and so on. Such an environment, of course, discouraged the evolution of trading links within the region. As urban populations grew and the number of plantation workers increased, the Europeans found it necessary to organise the distribution of food through governmental structures and to control both purchasing and retail prices.

PROCESS OF DE-COLONIALISATION

The process of de-colonialisation began after World War II and most African countries gained their independence in the 1950's and 1960s. African governments generally retained the marketing structures and trade barriers bequeathed to them by the colonial regime but they benefited from the relatively high commodity prices at the time. In the 1970s and early 80s some commodities, such as coffee, cocoa, sugar and rubber, were the subject of International Commodity Agreements which maintained prices at a level agreed by producing and consuming countries. Some African countries experimented with collectivised production as a means of improving the economies of scale and others concentrated on self-sufficiency and import substitution.

Many large commercial farms and plantations were retained and expanded. In the post World War II period, European countries embarked on programmes to develop their own agricultural sectors, partly in response to their experiences in the war, when large quantities of food had to be imported by sea at great cost both in money and lives. All developed countries were able to boost agricultural output through innovations in farming technology – machinery, artificial fertilizers, pesticides and new multiplication techniques. Farms too became bigger and more efficiently managed.

During this period, developed countries also enjoyed unprecedented rates of industrial and economic growth. Trade in manufactured goods and services increased enormously. Transport systems improved by air, sea and land and communications systems were developed to facilitate trade. Trade in raw materials and agricultural products has become a relatively minor component of international commerce. Farming now represents only about two percent of total economic output in the most highly developed countries.

Traditional African farming methods do not lend themselves so easily to efficient, large-scale production. Conflict Burundi, Eritrea, Ethiopia, Democratic Republic of the Congo, Rwanda, Sudan and Uganda have all suffered multiple problems from war or internal conflict. Their economies have been severely damaged by military expenditure, degradation of infrastructure, dislocation of populations, loss of labour to military activities and neglect of, and damage to, farms and factories.

Millions of ECA citizens have lost their lives over the last thirty years. Apart from the devastation of war itself, the disruption caused by conflict has delayed the implementation of economic reforms and development programmes and many of these countries have lost their access to aid programmes and investment opportunities. The reasons for conflict have included tribal animosity, territorial disputes, remnants of Cold War conflict, totalitarianism and religious intolerance.

Some analysts have identified a link between conflict and poverty in which it is difficult to break the cycle of poverty leading to unrest and unrest exacerbating poverty. At the end of 2001, however, there seem to be some grounds for optimism. Most major conflicts in the region have been resolved or, at least reduced to a lower level of intensity.

AGRICULTURAL DEVELOPMENT STRATEGIES

The lack of development in African countries has caused their economies to fall further and further behind those of the leading industrial nations. Many different development strategies have been tried. Some African countries have successfully encouraged investment in mining, tourism and industry. In agriculture, producers have been encouraged to move away from subsistence farming towards a more commercial approach as governments realised that income generated from the sale of surplus production could be used to improve productivity.

Agricultural development in ECA has faced an uphill struggle for the last twenty years. In an effort to stimulate development many countries borrowed heavily from bodies such as the IMF and from the commercial banking sector.

These loans were not granted without strings attached, however. Most African countries were obliged to liberalise their economies by adopting significant policy changes often applied in packages known as Structural Adjustment Programmes (SAPs).

These programmes included a number of elements but generally included requirement to:

- Devalue the currency (to discourage imports and make exports more competitive),
- To make the currency freely convertible with other currencies,
- To cut public expenditure (in order to lower taxes),
- To dismantle state controlled marketing boards,
- To privatise state-owned industries (to raise capital and stimulate competition),
- To cut import restrictions (to encourage local industries to become more efficient),
- To allow foreign companies to freely repatriate profits (to encourage inward investment),
- And to boost exports.

The economists who designed SAPs were convinced that the only way African countries could transform their economies was to encourage inward investment and earn foreign exchange to invest in infrastructure and lay the foundations for industrialisation.

These measures assumed that any country could compete in the world market if production and investment was concentrated in areas where they were deemed to have a competitive advantage. The only activity in which ECA nations could be said to have a competitive advantage in the world market was in the production of agricultural products and the exploitation of natural resources such as forestry, fishing and mining. The major flaw in this strategy was that similar advice was given to almost all tropical countries at the same time. Coffee-producing countries were encouraged to boost coffee production; sugar producers should produce more sugar, and so on. This resulted in over-production of these commodities which caused prices to plunge in the international markets. On average, current prices of tropical products (taking dollar inflation into account) are only about one seventh of those prevailing in 1980 (UN General Assembly). Economists call this phenomenon the fallacy of composition - less income is earned as more commodities are produced.

Another component of SAPs which many observers believe to have been counter-productive was the requirement to cut public expenditure. All too often this meant a cut in health programmes, education and agricultural extension. These measures have tended to reduce, rather than enhance the flexibility of the workforce and to curtail agricultural development.

Overall, the record of inward investment has been poor and the ending of currency controls has increased opportunities for transfer pricing abuse (where companies over-price imports and under-price exports to reduce tax liability). The most important SAP reform affecting the distri-bution of agricultural products has been the dismantling of state-controlled marketing boards and the practice of setting fixed purchasing and sales prices for commodities. It was assumed that government control of markets had obscured the forces of competition in supply and demand in the economy.

A free market system would unleash these forces and increase productivity. It would force producers to meet the demands of consumers both in price and quality.

Farmers would be able to buy inputs cheaper from competing suppliers, and the country, as a whole, would become more competitive in world markets. Unfortunately, competitive and transparent markets did not emerge spontaneously (Shepherd). Most African farmers have too little land to produce truck-loads of goods and they are widely dispersed over the countryside.

There is not enough business to encourage more than one trader to operate in many areas. Farmers have no means of communicating with the outside world or even the nearest town and they are often unwilling to risk the investment of bringing their goods to market resulting in considerable waste.

Laws may have been passed which ban collusion among traders to pay low prices to farmers and charge high prices to consumers, but there are often insufficient resources to enforce such laws. Most traders have no experience of free market conditions and are reluctant to put their fellow traders out of business with serious competition.

Advocates of SAPs point to examples of countries that have improved their economies after adopting SAPs (World Bank) but there are few in Africa. Most ECA countries were not able to implement SAPs until relatively recently but rates of poverty have increased in many of these countries. Intense conflict, both within and between countries of the region, drought, desertification and, now HIV/AIDS have further weakened economic development in ECA. Most critics of the reform process, however, acknowledge that markets in African countries must be made more competitive and SAPs are designed to do that but this process may take a considerable time.

Trade Agreements

Economic links between ECA countries and their former colonial rulers have been maintained since independence. The economies of these countries have been moulded to meet the needs of their European counterpart for a hundred years or more and it would have been difficult for them to make the necessary changes in production patterns to trade succe-ssfully with other countries.

The Europeans too needed to maintain supplies of raw materials and export markets in Africa and to protect the business of their trading companies. In 1975 all ten countries covered by this study became party to the Lomé Convention. The Convention established trade, aid and cultural relatio-nships between 15 European countries and 71 so called ACP (African, Caribbean and Pacific) countries which had either been colonies of, or had had strong historical links with, Europe. This agreement did not rule out bilateral or multilateral agreements with other countries but did give ACP countries preferential access to European markets.

ECA countries have also decided to try to stimulate regional trade by bringing their economies closer together in regional economic agreements such as COMESA and SADC. East Africans have exchanged goods and ideas with many other peoples of the world for millennia.

In these exchanges of goods, cultural links have been established which have influenced ECA life at all levels – in religion, the arts, public sector structures, the economy and agriculture. In the last decade or two, however, this process has accelerated tremendously.

There is no agreed definition of globalisation. It is simply a term which has been used recently to describe the impact of innovations in communication and transport systems on trade and the growing interdependence of nations due to economic sophistication and burgeoning output.

In addition, high levels of protection between trading blocks of countries are breaking down as barriers to trade are reduced. These changes have made it possible to increase the volume of trade between countries in agricultural products.

It became clear that overall levels of trade could be increased if trade barriers were reduced, where there was agreement to do so, and that international trade should be governed by mutually agreed rules. The most active trading nations have been keen to find new markets for their goods and to reduce the barriers to free trade. These countries, however, have been reluctant to expose their own markets to foreign competition, especially unfair competition from subsidised or sub-standard goods. At the international level, global liberalisation was stimulated by the General Agreement on Tariffs and Trade (GATT) which was first implemented in 1948 as a mechanism to promote free and fair trade among member countries.

Several rounds of negotiations of trade rules have occurred throughout the history of GATT. The Uruguay Round, which began in 1986, was the eighth of the GATT rounds. In April 1994, officials from more than 100 countries gathered in Marrakech, Morocco to sign the Uruguay Agreement and to confer the role of further trade reforms on the newly established World Trade Organisation. The reform process is by no means complete. Almost all countries have now committed themselves to the objectives associated with their membership of the WTO.

In order to meet these objectives, countries are obliged to further reform their existing internal economic and external trade policies. The future of trade and agriculture in ECA is inextricably linked to the rate and direction of these reforms.

7

Agricultural Economics

Agricultural economics originally applied the principles of economics to the production of crops and livestock — a discipline known as agronomics. Agronomics was a branch of economics that specifically dealt with land usage. It focused on maximizing the crop yield while maintaining a good soil ecosystem. Throughout the 20th century the discipline expanded and the current scope of the discipline is much broader. Agricultural economics today includes a variety of applied areas, having considerable overlap with conventional economics.

ORIGINS

Economics is the study of resource allocation under scarcity. Agronomics, or the application of economic methods to optimizing the decisions made by agricultural producers, grew to prominence around the turn of the 20th century. The field of agricultural economics can be traced out to works on land economics. Henry Charles Taylor was the greatest contributor with the establishment of the Department of Agricultural Economics at Wisconsin in 1909. Another contributor, Theodore Schultz was among the first to examine growth economics as a problem related directly to agriculture. Schultz was also instrumental in establishing econometrics as a tool for use in analysing agricultural economics empirically; he noted in his landmark 1956 article that agricultural supply analysis is rooted in "shifting sand," implying that it was and is simply not being done correctly.

Growth

One scholar summarizes the growth of agricultural economics as follows: "Agricultural economics arose in the late 19th century, combined the theory of the firm with marketing and organization theory, and developed throughout the 20th century largely as an empirical branch of general economics. The discipline was closely linked to empirical applications of mathematical statistics and made early and significant contributions to econometric methods. In the 1960's and afterwards, as agricultural sectors in the OECD countries contracted, agricultural economists were drawn to the growth problems of poor countries, to the trade and macroeconomic policy implications of agriculture in rich countries, and to a variety of production, consumption, and environmental and resource problems."

Agricultural economists have made many well-known contributions to the economics field with such models as the cobweb model, hedonic regression pricing models, new technology and diffusion models (Zvi Griliches), multifactor productivity and efficiency theory and measurement, and the random coefficients regression. The farm sector is frequently cited as a prime example of the perfect competition economic paradigm.

Since the 1970s, agricultural economics has primarily focused on seven main topics, according to a scholar in the field: agricultural environment and resources; risk and uncertainty; consumption and food supply chains; prices and incomes; market structures; trade and growth; and technical change and human capital;.

In terms of technical change, there have been increasingly rapid growths and innovations in the equipment designed for agricultural research. This equipment includes instruments for plant physiology research, and monitoring soil conditions and atmospheres.

ECONOMIC DEVELOPMENT

Economic development, the process whereby simple, low-income national economies are transformed into modern industrial economies. Although the term is sometimes used as a synonym for economic growth, generally it is employed to describe a change in a country's economy involving qualitative as well as quantitative improvements. The theory of economic development—how primitive and poor economies can evolve into sophisticated and relatively prosperous ones—is of critical importance to underdeveloped countries, and it is usually in this context that the issues of economic development are discussed.

Economic development first became a major concern after World War II. As the era of European colonialism ended, many former colonies and other countries with low living standards came to be termed underdeveloped countries, to contrast their economies with those of the developed countries, which were understood to be Canada, the United States, those of western Europe, most eastern European countries, the then Soviet Union, Japan, South Africa, Australia, and New Zealand. As living standards in most poor countries began to rise in subsequent decades, they were renamed the developing countries.

There is no universally accepted definition of what a developing country is; neither is there one of what constitutes the process of economic development. Developing countries are usually categorised by a per capita income criterion, and economic development is usually thought to occur as per capita incomes rise. A country's per capita income (which is almost synonymous with per capita output) is the best available measure of the value of the goods and services available, per person, to the society per year. Although there are a number of problems of measurement of both the level of per capita income and its rate of growth, these two indicators are the best available to provide estimates of the level of economic well-being within a country and of its economic growth.

It is well to consider some of the statistical and conceptual difficulties of using the conventional criterion of underdevelopment before analysing the causes of underdevelopment. The statistical difficulties are well known. To begin with, there are the awkward borderline cases. Even if analysis is confined to the underdeveloped and developing countries in Asia, Africa, and Latin America, there are rich oil countries that have per capita incomes well above the rest but that are otherwise underdeveloped in their general economic characteristics.

Second, there are a number of technical difficulties that make the per capita incomes of many underdeveloped countries (expressed in terms of an international currency, such as the U.S., dollar) a very crude measure of their per capita real income. These difficulties include the defectiveness of the basic national income and population statistics, the inappropriateness of the official exchange rates at which the national incomes in terms of the respective domestic currencies are converted into the common denominator of the U.S., dollar, and the problems of estimating the value of the non-cash components of real incomes in the underdeveloped countries. Finally, there are conceptual problems in interpreting the meaning of the international differences in the per capita income levels.

Although the difficulties with income measures are well established, measures of per capita income correlate reasonably well with other measures of economic well-being, such as life expectancy, infant mortality rates, and literacy rates. Other indicators, such as nutritional status and the per capita availability of hospital beds, physicians, and teachers, are also closely related to per capita income levels. While a difference of, say, 10 percent in per capita incomes between two countries would not be regarded as necessarily indicative of a difference in living standards between them, actual observed differences are of a much larger magnitude. India's per capita income, for example, was estimated at $270 in 1985. In contrast, Brazil's was estimated to be $1,640, and Italy's was $6,520. While economists have cited a number of reasons why the implication that Italy's living standard was 24 times greater than India's might be biased upward, no one would doubt that the Italian living standard was significantly higher than that of Brazil, which in turn was higher than India's by a wide margin.

The interpretation of a low per capita income level as an index of poverty in a material sense may be accepted with two qualifications. First, the level of material living depends not on per capita income as such but on per capita consumption. The two may differ considerably when a large proportion of the national income is diverted from consumption to other purposes; for example, through a policy of forced saving. Second, the poverty of a country is more faithfully reflected by the representative standard of living of the great mass of its people. This may be well below the simple arithmetic average of per capita income or consumption when national income is very unequally distributed and there is a wide gap in the standard of living between the rich and the poor.

The usual definition of a developing country is that adopted by the World Bank: "low-income developing countries" in 1985 were defined as those with per capita incomes below $400; "middle-income developing countries" were defined as those with per capita incomes between $400 and $4,000. To be sure, countries with the same per capita income may not otherwise resemble one another: some countries may derive much of their incomes from capital-intensive enterprises, such as the extraction of oil, whereas other countries with similar per capita incomes may have more numerous and more productive uses of their labour force to compensate for the absence of wealth in resources. Kuwait, for example, was estimated to have a per capita income of $14,480 in 1985, but 50 percent of that income originated from oil.

In most regards, Kuwait's economic and social indicators fell well below what other countries with similar per capita incomes had achieved. Centrally planned economies are also generally regarded as a separate class, although China and North Korea are universally considered developing countries. A major difficulty is that prices serve less as indicators of relative scarcity in centrally planned economies and hence, are less reliable as indicators of the per capita availability of goods and services than in market-oriented economies.

Estimates of percentage increases in real per capita income are subject to a somewhat smaller margin of error than are estimates of income levels. While year-to-year changes in per capita income are heavily influenced by such factors as weather (which affects agricultural output, a large component of income in most developing countries), a country's terms of trade, and other factors, growth rates of per capita income over periods of a decade or more are strongly indicative of the rate at which average economic well-being has increased in a country.

ECONOMIC DEVELOPMENT AS AN OBJECTIVE OF POLICY

Motives for Development

The field of development economics is concerned with the causes of underdevelopment and with policies that may accelerate the rate of growth of per capita income. While these two concerns are related to each other, it is possible to devise policies that are likely to accelerate growth (through, for example, an analysis of the experiences of other developing countries) without fully understanding the causes of underdevelopment. Studies of both the causes of underdevelopment and of policies and actions that may accelerate development are undertaken for a variety of reasons. There are those who are concerned with the developing countries on humanitarian grounds; that is, with the problem of helping the people of these countries to attain certain minimum material standards of living in terms of such factors as food, clothing, shelter, and nutrition. For them, low per capita income is the measure of the problem of poverty in a material sense.

The aim of economic development is to improve the material standards of living by raising the absolute level of per capita incomes. Raising per capita incomes is also a stated objective of policy of the governments of all developing countries. For policy-makers and economists attempting to achieve their governments' objectives, therefore, an understanding of economic development, especially in its policy dimensions, is important. Finally, there are those who are concerned with economic development either because they believe it is what people in developing countries want or because they believe that political stability can be assured only with satisfactory rates of economic growth. These motives are not mutually exclusive. Since, World War II many industrial countries have extended foreign aid to developing countries for a combination of humanitarian and political reasons.

Those who are concerned with political stability tend to see the low per capita incomes of the developing countries in relative terms; that is, in relation to the high per capita incomes of the developed countries. For them, even if a developing country is able to improve its material standards of living through a rise in the level of its per capita income, it may still be faced with the more intractable subjective problem of the discontent created by the widening gap in the relative levels between itself and the richer countries.

(This effect arises simply from the operation of the arithmetic of growth on the large initial gap between the income levels of the developed and the underdeveloped countries. As an example, an underdeveloped country with a per capita income of $100 and a developed country with a per capita income of $1,000 may be considered. The initial gap in their incomes is $900. Let the incomes in both countries grow at 5 percent. After one year, the income of the underdeveloped country is $105, and the income of the developed country is $1,050. The gap has widened to $945. The income of the underdeveloped country would have to grow by 50 percent to maintain the same absolute gap of $900.) Although there was once in development economics a debate as to whether raising living standards or reducing the relative gap in living standards was the true desideratum of policy, experience during the 1960–80 period convinced most observers that developing countries could, with appropriate policies, achieve sufficiently high rates of growth both to raise their living standards fairly rapidly and to begin closing the gap.

DETERMINING ECONOMIC VALUES

Once financial prices or costs and benefits have been determined and entered in the project accounts, the analyst estimates the economic value of a proposed project to the nation as a whole. The financial prices are the starting point for the economic analysis; they are adjusted as needed to reflect the value to the society as a whole of both the inputs and outputs of the project.

When the market price of any good or service is changed to make it more closely represent the opportunity cost (the value of a good or service in its next best alternative use) to the society, the new value assigned becomes the "shadow price" (sometimes referred to as an "accounting price"). In the strictest sense, a shadow price is any price that is not a market price, but the term usually also carries the connotation that it is an estimate of the economic value of the good or service in question, perhaps weighted to reflect income distribution and savings objectives.

The purposes of project analysis, we took the objective of a farm to be to maximize the farm family's incremental net benefit, the objective of the firm to maximize its incremental net income, and the objective of the society to maximize the contribution a project makes to the national income-the value of all final goods and services produced in the country during a particular period. These objectives, and the analysis to test their realisation, were seen in financial terms for farms and firms. But economic analysis of a project moves beyond financial accounting.

Strictly speaking, we may say that in financial analysis our numeraire-the common yardstick of account-is the real income change of the entity being analysed valued in domestic market prices and in general expressed in domestic currency. But in economic analysis, since, market prices do not always reflect scarcity values, our numeraire becomes the real, net national income change valued in opportunity cost. As One methodology expresses these economic values in domestic currency and uses a shadow price of Foreign Exchange; the shadow price increases the value of traded goods to allow for the premium on foreign exchange arising from distortions caused by trade policies. Another method in use expresses the opportunity cost value of real national income change in domestic currency converted from foreign exchange at the official exchange rate and applies a conversion factor to the opportunity cost or value in use of non-traded goods expressed in domestic currency; the conversion factor reduces the value of non-traded goods relative to traded goods to allow for the foreign exchange premium.

The adjusting financial accounts to reflect economic values commences, an important practical consideration must be emphasized. Many of the adjustments to the financial accounts can become quite complex. To every agricultural project, nor will all points have the same importance in those projects where they do apply.

The complexity of some calculations and the relative importance of some adjustments recall the reason for undertaking an economic analysis of a project: to improve the Investment decision. Some adjustments will make a considerable difference to the economic attractiveness of a proposed project; others will be of minor importance, and no reasonable adjustment would change the Investment decision. What we need to do here is to adopt an accounting practice-the doctrine of materiality. The analyst must focus his attention on those adjustments to the financial accounts that are likely to make a difference in the project Investment decision. He should use rough approximations or ignore trivial adjustments that will not make any difference in the decision. There is an important balance to be struck between analytical elegance and getting on with the job.

We will adjust the financial prices of tangible items to reflect economic values in three successive steps: (1) adjustment for direct transfer payments, (2) adjustment for price distortions in traded items, and (3) adjustment for price distortions in non-traded items. Before embarking on this series of adjustments, we will examine the problem of determining the appropriate premium for Foreign Exchange. After completing the adjustments, we will summarise the main points in a "decision tree" for determining economic values.

The series of successive adjustments to the financial accounts will lead to a set of economic accounts in which all values are stated in "efficiency prices," that is, in prices that reflect real resource use or consumption satisfaction and that are adjusted to eliminate direct and indirect transfers. These values will be market prices when market prices are good estimates of economic value or they will be shadow prices when market prices have had to be adjusted for distortions. When we adjust financial prices to reflect economic values better, in the vast majority of cases we will use the opportunity cost of the good or service as the criterion.

We will use opportunity costs to value all inputs and outputs that are intermediate products used in the production of some other good or service. For some final goods and services, however, the concept of opportunity cost is not applicable because it is consumption value that sets the economic value, not value in some alternative use. In these instances, we will adopt the criterion of "willingness to pay" (also called "value in use"). We need to do this, however, only when the good or service in question is non-traded (perhaps as a result of government regulation) during some part of the life of the project-a point. Because the ultimate objective of all economic activity is to satisfy consumption wants, all opportunity costs are derived from consumption values, and thus from willingness to pay.

An example may clarify our use of willingness to pay and opportunity cost. Suppose a country that is a rather inefficient producer of sugar has a policy to forbid sugar imports to protect its local industry. The price of sugar may then rise well above what it would be if sugar were imported. Even at these higher prices, most consumers will still buy some sugar for direct consumption-say, in coffee or tea-even though they may use less sugar than if the price were lower.

The domestic price of sugar will be above the world market price and will represent the value of the sugar by the criterion of willingness to pay. If we were now to consider the economic value of sugar from the standpoint of its use in making fruit preserves, its value would become the opportunity cost of diverting the sugar from direct consumption, where willingness to pay is the criterion and has set the economic value.

Economic analysis, then, will state the cost and benefit to the society of the proposed project investment either in opportunity cost or in values determined by the willingness to pay. The costs or values will be determined in part by both the resource constraints and the policy constraints faced by the project. The difference between the benefit and the cost-the incremental net benefit stream-will be an accurate reflection of the project's income-generating capacity-that is, its net contribution to real national income.

The system outlined here will make no adjustment for the income distribution effects of a proposed project nor for its effect on the amount of the benefit generated that will be Invested to accelerate future growth. Rather, the economic project analysis, stated in efficiency prices, will judge the capacity of the project to generate national income. The analyst can then choose from those alternative projects (or alternative formulations of roughly the same project) the high-yielding alternative that in his subjective judgement also makes the most effective contribution to objectives other than maximizing national income-objectives such as income distribution, savings generated, number of jobs produced, regional development, national security, or whatever. The choice about the *kind* of project will of course be made rather early in the project cycle.

Thus, it may be determined early on that for reasons of social policy a project will be preferred that encourages smallholder agriculture rather than plantations. Then, the choices will likely be several projects or variants of projects that encourage smallholders; the analytical technique presented here can determine from among the projects that will further the desired social objective the ones that are more economically efficient.

Although the system outlined here makes no adjustment for income distribution effects or for saving versus consumption, it is compatible with other analytical systems that do. In particular, Squire and van der Tak *(1975)* recommend evaluating proposed projects first by using essentially the same efficiency prices that will be estimated here and then by further adjusting these prices to weight them for income distribution effects and for potential effects on further investment of the benefits generated. Making allowances for income distribution and savings effects involves somewhat more complex adjustments than those necessary to estimate efficiency prices; it also unavoidably incorporates some element of subjective judgement. Although these systems have attracted widespread interest among economists, their application has been only partial or on a limited scale. The system of economic analysis using efficiency prices that is outlined here is essentially the one currently used for all but a few World Bank projects and also the one used for most analyses of projects funded by other international organisations.

In the economic analysis, we will want to work with accounts cast on a constant basis; thus we will want to be sure that any inflation contingency allowances have been taken out. However, physical contingency allowances and contingency allowances intended to allow for *relative* price changes are properly incorporated in the economic accounts, even when the accounts are in constant prices. Of course, any of the items included among the contingencies may be revalued, if necessary, to adjust them from their market prices to economic values.

The projected financial accounts will usually not have any entry for cash. Instead, they will show separately the cash position of the farmer or note a cumulative cash surplus or deficit. It is possible, however, that some accounts may have a cash balance included in an entry for working capital or the like. If such an entry exists, it must be removed from the economic analysis; since, we will be working on a real basis in the economic accounts, we will show real costs when they occur and real benefits when they are realised.

DETERMINING THE PREMIUM ON FOREIGN EXCHANGE

Adjusting the financial accounts of a project to reflect economic values involves determining the proper premium to attach to foreign exchange. That determination quickly involves issues of obtaining proper values and of economic theory. Fortunately for most agricultural project analysts, the answer to the question about how to determine the foreign exchange premium is simple (and simplistic): ask the central planning agency.

The point is that if various alternative investment opportunities open to a nation are to be compared, the same foreign exchange premium must be used in the economic analysis of each alternative. Otherwise we will be mixing apples and oranges and cannot use our analysis reliably to choose among alternatives. Sometimes, however, the analyst will be forced to make his own estimate of the foreign exchange premium.

The need to determine the foreign exchange premium arises because in many countries, as a result of national trade policies (including tariffs on imported goods and subsidies on exports), people pay a premium on traded goods over what they pay for non-traded goods. This premium is not adequately reflected when the prices of traded goods are converted to the domestic currency equivalent at the official exchange rate.

The premium represents the additional amount that users of traded goods, on an average and throughout the economy, are willing to pay to obtain one more unit of traded goods. Since, all costs and benefits in economic analysis are valued on the basis of opportunity cost or willingness to pay, it is the relation between willingness to pay for traded as opposed to non-traded goods that establishes their relative value.

The premium people are willing to pay for traded goods, then, represents the amount that, on the average, traded goods are mispriced in relation to non-traded items when the official exchange rate is used to convert foreign exchange prices into domestic values.

By applying the premium to traded goods, we are able to compare the values of traded and non-traded goods by the criterion of opportunity cost or willingness to pay. Although this premium is commonly referred to as the foreign exchange premium, it should be recognised that the premium is actually a premium for traded goods; foreign exchange itself has no intrinsic value. The premium for traded goods is a premium on the particular "basket" of traded goods that the present and projected trade pattern implies.

Of course, future patterns of trade could change the exact composition of the basket, and thus the premium would change; to estimate these changes involves a knowledge of elasticities-the way demand and supply of goods and services vary when prices change-that is generally not available. Where such elasticities are known, it is possible for a well-trained economist to provide the project analyst with a more accurate estimate of the expected premium on foreign exchange.

If traded items were to be taken into the project analysis at an economic value obtained by simply multiplying the border price by the official exchange rate without adjusting for the foreign exchange premium, imported items would appear too cheap and domestic items too dear. This would encourage overinvestment in projects that use imports. For example, if combine harvesters look cheap because no allowance is made for the premium on traded goods, then imported combines might displace local harvest labour, even though the local labour might have no other opportunities for employment.

There are two equivalent ways of incorporating the premium on foreign exchange in our economic analysis. The first is to multiply the official exchange rate by the foreign exchange premium, which yields a shadow foreign exchange rate. [Note that this derivation of the shadow exchange rate is appropriate for efficiency analysis of projects and thus has a discrete definition. Other definitions of the shadow exchange rate are appropriate depending on the uses to which the rate will be put. Reesearchers (1972) discuss some of these alternatives.]

The shadow exchange rate is then used to convert the foreign exchange price of traded items into domestic currency. The effect of using the shadow exchange rate is to make traded items relatively more expensive in domestic currency by the amount of the foreign exchange premium. (An alternative arithmetic formulation is to convert the foreign exchange price into domestic currency at the official exchange rate and then multiply by 1 plus the foreign exchange premium stated in decimal terms.) The shadow exchange rate approach has been used in the past in most World Bank projects when adjustments have been made to allow for the foreign exchange premium on traded goods, and it is also used in the UNIDO Guidelines.

An alternative way to allow for the foreign exchange premium on traded items that is increasingly coming into use is to reduce the domestic currency values for non-traded items by an amount sufficient to reflect the premium. This is sometimes called the "conversion factor" approach. In its simplest form, based on straightforward efficiency prices, a single conversion factor - the "standard conversion factor" of Squire and van der Tak - is derived by taking the ratio of the value of all exports and imports at border prices to their value at domestic prices.

In this form, the standard conversion factor bears a close relation to our shadow exchange rate; indeed, the standard conversion factor may be determined by dividing the official exchange rate by the shadow exchange rate or by taking the reciprocal of 1 plus the foreign exchange premium stated in decimal terms. Market prices or shadow prices of non-traded items are then multiplied by this standard conversion factor, and this reduces them to their appropriate economic value.

Little and Mirrlees and Squire and van der Tak both adopt the conversion factor approach. In addition, both pairs of authors recommend deriving specific conversion factors for particular groups of products that will allow for any difference between market prices and opportunity costs and for the foreign exchange premium on traded items. As a result, their specific conversion factors may always be applied directly to domestic market prices. These authors also recommend that their conversion factors be calculated in social prices by including distribution weights.In the valuation system followed here, all items are valued at efficiency prices without allowance for distribution weights (the issue of selecting projects to achieve distributional objectives is treated as a subsequent decision). This being the case, consideration of the distribution-weighted conversion factors proposed by Little and Mirrlees and Squire and van der Tak may be left aside, and we may focus our discussion on the Squire and van der Tak standard conversion factor as it relates to efficiency prices.

The relation between the official exchange rate, the foreign exchange premium (Fx premium), the shadow exchange rate (SER), and the standard conversion factor (SCF) is perhaps easier to understand in equation form:

$$\text{OER} \times (1 + \text{Premium}) = \text{SER and } \frac{1}{(1{+}\text{FX premum})} = \text{SCF}$$

so that, as Squire and van der Tak note, and We may illustrate these relations

by an example taken from the Agricultural Minimum Package Project in Ethiopia. At the time the project was appraised, the analyst knew that the official exchange rate of Eth$2.07 = US$1 failed to account for a foreign exchange premium of at least 10 percent. (The symbol for the Ethiopian dollar is Eth$; since, this project was appraised the name of the currency unit has been changed to birr.) Thus, the analyst multiplied the official exchange rate by 1 plus a 10 percent foreign exchange premium to obtain a shadow exchange rate of Eth$2.28 = US$1 (2.07 × 1.1 = 2.28) that he rounded up to Eth$2.30 = US$ 1. The shadow exchange rate was then applied to all'traded items in the financial accounts, thereby increasing their relative value.

If the domestic currency is worth more per unit than the foreign exchange, the arithmetic is somewhat different. At the time the Nucleus Estate/ Smallholder Oil Palm Project in Rivers State, Nigeria, was appraised, the official exchange rate was N 1 = US$1.54. (The symbol for Nigerian naira is N.) The project analysts were given a shadow exchange rate of N1 = US$1.27 to use in their economic evaluation. If, however, they had simply been informed that the foreign exchange premium was 21 percent, they could have determined the shadow exchange rate by dividing the dollar value by 1 plus the premium stated in decimal terms (1.54 - 1.21 = 1.27).

Of course, the effect of applying the shadow exchange rate to the traded items in the Ethiopian project was to make all non-traded items 10 percent less expensive in relation to the traded items in the economic accounts as opposed to the financial accounts.

Now, instead of increasing the relative value of traded items, we could reduce the value of all non-traded items appearing in the financial accounts so that in the economic account they are relatively 10 percent less expensive. To do this we calculate the standard conversion factor, which is 1 divided by 1 plus the amount of the foreign exchange premium stated in decimal terms.

In this case, the result is a factor of 0.909 (1 - 1.1 = 0.909). To obtain the economic values, we would then multiply all financial prices for non-traded items by this factor if these market prices have been judged good estimates of opportunity cost or good estimates of economic value on grounds of willingness to pay. For non-traded items such as wage rates for unskilled labour for which it is felt that the market price has overstated the economic values, we would first determine a good estimate of the economic value in domestic currency and then multiply that by the standard conversion factor. Financial prices for traded items, whether imports or exports, would be left unchanged in the economic accounts except that any transfer payment included in these prices would be taken out. To get all values into the same currency, we would convert all foreign currency prices to domestic currency values using the official exchange rate.

When we turn to determining measures of project worth, we will find that the absolute value of the net present worth differs depending on which approach we use, shadow exchange rate or conversion factor, but that the relative net present worths of different projects analysed by the same approach will not change. Whichever approach is used, the internal rate of return, the benefit-cost ratio, and the net benefit-investment ratio do not change. (Using a number of disaggregated conversion factors, rather than a standard conversion factor, can give different values for the measures of project worth. Hence, for projects at the margin of acceptability, using specific conversion factors rather than a standard conversion factor or a shadow exchange rate may result in a different decision on whether to accept or reject, but such cases are infrequent.)

ROLE AND IMPORTANCE OF AGRICULTURE IN INDIAN ECONOMY

Farming is the backbone of the Indian economy. Despite major emphasis on industrial development during the last four decades, farming continues to occupy a place of pride in our economy. The importance of farming can be brought out from the following facts:

SHARE IN NATIONAL EARNINGS

Although the share of farming in the total national earnings has been gradually decreasing on account of the development of the secondary and tertiary sectors, it still contributed about 18 per cent of nation earnings in 2006-07. (In 1950-51, it was 59 per cent)

SOURCE OF EMPLOYMENT

In India, farming is the main source of employment. Even in 2004-05, more than 56 per cent of the total labour force of India is engaged in farming and depend on it for their livelihood (1950-51: 69.5 per cent).

It becomes evident from this fact that other sectors of the economy could not generate enough employment for the growing population.

PROVISION OF FOOD GRAINS

In a developing country like India where a very large proportion of earningsare spent on food and the population is increasing rapidly, the demand for foodgrains has been increasing at a fast rate. Farming In India has played an important role in meeting almost the entire food needs of the people. The production of foodgrains in India has increased from 51 million tones in 1950-51 to 208.3 million tones in 2005-06, *i.e.* By a little more than 4 times since 1950-51. This has enabled the country to overcome the problems of foodgrain shortages. The country is almost self-sufficient in food grains and it no longer depends on import of food grains.

SUPPLY OF RAW MATERIALS TO INDUSTRIAL SECTOR

Farming plays an important role in industrial development. Many industries like cotton industry, jute industries, sugar industries, food processing industries, etc. Depends on farming for their raw material requirements. Moreover, workers engaged in various industries depend on farming for their food requirement.

MARKET FOR INDUSTRIAL PRODUCT

Farming provides markets for a large number of industrial products. Since about two thirds of India lives in rural areas, there has been a large rural purchasing power which has created a large demand for all types of industrial products. Green revolution has considerably increased the purchasing power of the large farmers substantially in the recent years. Thus for the demand for various products like soaps, detergents, clothes, cycles, scooters, radios, television, torches, lead batteries, etc. Has witnessed a marked increase. Likewise, the demand for a variety of farm inputs like chemical fertilizer, tractors, pump-sets, pesticides etc. Has increased sharply. This has stimulated the development of industries producing these inputs.

EARNER OF FOREIGN EXCHANGE

Farming plays an important role in the Indian economy as an earner of foreign exchange through exports of farm commodities like tea, cotton, coffee, jute, fruits, vegetables, spices, tobacco, sugar, oil, cashew kernels, etc. In the past, export of farm products accounted for about 70 per cent of the export earnings of the country. However, with economic development and consequent diversification of our exports, the share of farming in total exports has come down to about 10 per cent in 2005-06. All these exports bring valuable foreign exchange to pay for the increased imports of machinery and raw materials required in the non- farm sector.

SIGNIFICANCE FOR TRADE AND TRANSPORT

Farming helps in the development of tertiary (or service) sector. For example various means of transport like roadways and railways get the bulk of their business from the movement of farm commodities and raw materials. A significant part of internal trade constitutes mainly of farm products.

SOURCE OF REVENUE FOR THE GOVERNMENT

Through the direct contribution of farm taxes to the central and state governments is not significant, they get a significant part of their total revenue in terms of land revenue, irrigation charges, taxes imposed on the commodities purchased by the farmers etc. The central government also earns revenue from export duties on farm production. Freight charges imposed by Indian Railways for carrying farm products generate huge revenue to the central exchequer. On over all view, India has always been benefited by farming. Though the future of India is industrialisation, the contribution of farming would always prove to be vital for making India a powerful and stable economy in the future.

IMPORTANCE OF FARMING

Farming, for decades, had been associated with the production of basic food harvests. Farming and farmingwere synonymous so long as farming was not commercialised. But as the process of economic development accelerated, many more other occupations allied to farming came to be recognised as a part of agriculture.

At present, farming and agriculture besides include forestry, fruit cultivation, dairy, poultry, mushroom, bee keeping, arbitrary, etc. Today, marketing, processing, distribution of farm products etc. Are all accepted as a part of modern agriculture. Thus, agriculture may be defined as the production, processing, marketing and distribution of harvests and livestock products. According to Webster's Dictionary, "farming is the art or science of production of harvests and livestock on farm."

IMPORTANCE OF AGRICULTURE

India is an farm country. The Indian economy is basically agrarian. In spite of economic development and industrialisation, farming is the backbone of the Indian economy. As Mahatma Gandhi said, "India lives in villages and farming is the soul of Indian economy".

Nearly two-thirds of its population depend directly on farming for its livelihood. Farming is the mainstay of India's economy. It contributes about 26 percent of the gross domestic product. Farming meets food requirements of the people and produces several raw materials for industries.

From the farm point of view, India is a unique country. It has a vast expanse of level land, rich soils, wild climatic variations suited for various types of harvests, ample sunshine and a long growing season. The net sown area in India today is about 143 million hectares. India has the highest percentage of land under cultivation in the world. In spite of the fact that large areas in India, after independence, have been brought under irrigation, only one-third of the harvested area is actually irrigated. The productivity of farming is very low. Farming depends mainly upon monsoon rain.

Most of the production comprises food harvests. About one-third of the land holdings is small, less than one hectare in size. Farmers on their own small prices of land and grow harvests primarily for consumption. Even storage facilities for harvests are inadequate. Now use of pesticides and fertilizers has increased and large areas have been brought under a high yielding variety of seeds. This led to the green revolution in several parts of India. This has helped in increasing yields per hectare as well as total production of different harvests. There are many reasons responsible for the low productivity of farming. About one-third of land holdings is very small less than one hectare in size.

Due to the small size of land holdings we cannot use modern way of cultivation. Even today the farmers are using very old methods, tools and implements for farming. Farmers are not using artificial ways of cultivation. Inputs like-better quality of seeds, fertilizers and pesticides are also not used by most of the farmers. The injustice of marginal farmers is also responsible. There is also a low productivity because of increasing pressure on land and absence of bank credit.

Farming is the backbone of our Indian economy. Farm development is a precondition of our national prosperity. It is the main source of earning livelihood of the people. Nearly two-thirds of its population depend directly on farming. Farming provides direct employment to 70 percent of working people in the country. It is the mainstay of India's economy.

Apart from those who are directly involved in the agrarian sector, a large number of the population is also engaged in agro-based activities. Farming meets the food requirements of large population of India. It ensures food security for the country. Substantial increase in the production of food grain like-rice, wheat etc. And non-food grains like-tea, coffee, spices, fruits and vegetables, sugar, cotton etc. Has made India self-sufficient. Farming also contributes to the national earnings of our country. It accounts for 26 percent of the gross domestic product.

The growth of most of the industries depends on farming. It produces several materials for industries. It forms the basis of many industries of India like-cotton, textile, jute, sugar industries etc.By providing cotton, sugarcane, oilseeds etc. People engaged in farming also buy the products of industries like-tractors, pesticides, fertilizers, pump-set etc. Farming contributes in foreign exchange of our country. India exports farm products like tea, coffee, sugar, tobacco, spices etc.And earns foreign currency. Exports from the farm sector have helped India in earning valuable foreign exchange and thereby boosting economic development. From above mentioned facts it is very clear that in spite of industrial development still farming is the backbone of the Indian economy.

The Five-year Plans accorded priority to the farm sector. In the past 50 years the food grain production in the country increased substantially from 51 million tonnes in 1950-51 to 209 million tonnes in 1999-2000. In spite of the constant rise of population, we have been able to build a good stock of 44.7 million tonnes in 2001. This is because of the scientific and institutional reforms in our country. The Indian government took several steps to improve the farm condition in the country. The government has encouraged consolidation of land holdings to promote the use of modern farm machines. Land reforms were introduced. The government took the lands of big land owners away and redistributed to landless laborers.

The government abolished the Zamindari Method. Modern methods of cultivation were introduced in the country. The government provided better infrastructure facilities such as— irrigation, electricity and transportation. Farm equipments such as— tractors, pump harvesters, fertilizers, pesticides were made available to farmers. Getting finance from banks was made easier for the farmers. The harvest insurance was another step to protect the farmers against losses caused by the harvest failure on account of natural calamities like drought, flood cyclone etc. High-yielding varieties of seeds, fertilizers and irrigation gave birth to Green Revolution. All these led to a tremendous increase in the production of harvests.

The country on an average, has enough in stock to meet the food requirements of its citizens. India has emerged as the largest producer of coconut, ginger, cashew nut, black-peeper and as the second largest producers of fruits and vegetables. The productivity of the land has increased through the years, but has not reached international productivity levels. Indian farming has diversified into various sectors and contributes significantly to the nation's economy. But this situation is not likely to remain so easy in the years to come. The population of India is likely to be around 1300 million. This would require a huge amount of food grains along with non-food grains. India has to use its vast potential of farming in a method etc.And planned manner. We have developed some of the techniques which the developed countries have been using.

Source of Livelihood:

In India the main occupation of our working population is farming. About 70 per cent of our population is directly engaged in farming. In advanced countries, this ratio is very small being 5 per cent in U.K., 4 per cent in the USA., 16 per cent in Australia, 14 per cent in France, 21 per cent in Japan and 32 per cent in USSR. This high proportion in farming is due to the fact that the non-farm activities have not been developed to absorb the rapidly growing population.

Contribution to National Earnings:

Framing is the premier source of our national earnings. According to National Earnings Committee and C.S.O., in 1960-61, 52 per cent of national earnings was contributed by farming and allied occupations. In 1976-77, this sector alone contributed 42.2 per cent while in 1981-82, its contribution was to the tune of 41.8 per cent. In 2001-02, it contributed around 32.4 per cent of national earnings. This was further reduced to 28 per cent in 1999-2000. Contrary to this, the proportion of farming in the U.K. is only 3.1, in the USA it is 3 percent, 2.5 per cent in Canada, 6 per cent in Japan, 7.6 per cent in Australia. The mere conclusion of all this is that more developed a country the smaller is the contribution of farming in national output.

Supply of Food and Fodder:

Farming sector also provides fodder for livestock (35.33 cores). Cow and buffalo provide protective food in the form of milk and they also provide draught power for farm operations. Moreover, it also meets the food requirements of the people. Import of food grains has been very small in recent years, rather export avenues are being looked for.

Importance in International Trade:

It is the farm sector that feeds the country's trade. Farm products like tea, sugar, rice, tobacco, spices etc. Constitute the main items of exports of India. If the development process of farming is smooth, export increases and imports are reduced considerably. Thus, it helps to reduce the adverse balance of payments and save our foreign exchange. This amount can be well utilised to import other necessary inputs, raw-material, machinery and other infrastructure which is otherwise useful for the promotion of economic development of the country.

MARKETABLE SURPLUS:

The development of farm sector leads to marketable surplus. As a country develops more and more people are to be engaged in mining, manufacturing and other non- farm sector. All these people depend upon the food production which they can meet from the marketable surplus. As farm development takes place, output increases and marketable surplus expand. This can be sold to other countries. Here, it is worth mentioning that the development of Japan and other countries were made possible by the surplus of farming. There is no reason why this could not be done in our own case.

Source of Raw Material:

Farming has been the source of raw materials to the leading industries like cotton and jute textiles, sugar, tobacco, edible and non-edible oils etc. All these depend directly on farming. Apart from this, many others like processing of fruits and vegetables, dal milling, rice husking, gur making also depend on farming for their raw material. According to United Nations Survey, the industries with raw material of farm origin accounted for 50 per cent of the value added and 64 percent of all jobs in the industrial sector.

Importance in Transport:

Farming is the main support for railways and roadways which transport bulk of farm produce from the farm to the man dies and factories. Internal trade is mostly in farm products. Besides, the finance of the govt, also, to the largest extent, depends upon the prosperity of the farm sector.

Contribute to Foreign Exchange Resources:

Farm sector constitutes an important place in the country's export trade. According to an estimate, farm commodities like jute, tobacco, oilseeds, spices, raw cotton, tea and coffee accounted for about 18 per cent of the total value of exports in India. This shows that farming products still continue to be a significant source of earning foreign exchange.

Vast Employment Opportunities:

The farm sector is significant as it provides greater employment opportunities in the construction of irrigation projects, drainage method and other such activities. With the fast growing population and high incidence of unemployment and disguised unemployment in backward countries, it is the only farming sector which provides more employment chances to the labour force. In this way, significance of farming emerges more and more.

Overall Economic Development:

In the course of economic development, farming employs the majority of people. This means raising the level of the national earnings and standard of living of the common man. The rapid" rate of growth in farming sector gives a progressive outlook and further motivation for development. As a result, it helps to create proper atmosphere for general economic development of the economy. Thus, economic development depends on the rate at which farming grows.

Source of Saving:

Progress in farming can go a long way in increasing savings. It is seen that rich farmers have started saving especially after the green revolution in the country. This surplus amount can be invested in the farming sector for further; development of the sector. Saving potentials are large in farming sector which can be properly tapped for the development of the country.

Source of Government Earnings:

In India, many state governments get sizeable revenue from the farming sector. Land revenue, farmearnings tax, irrigation tax and some other types of taxes are being levied on farming by the state governments. Moreover, considerable revenue is earned by way of excise duty and export duty on farm products.

Raj committee on Farm Taxation has suggested the imposition of taxation on farmearnings for raising revenue.

Basis of Economic Development:

Prof. Nurkse has laid sufficient emphasis on the progress of farming for a balanced growth of an economy. The development of farming provides the necessary capital for the development of other sectors like industry, transport and foreign trade. In fact, a balanced development of farming and industry is the need of the day. From the above explanation, it may be concluded that farming occupies an important place in the development of an economy. It is in fact, a pre-condition for economic upliftment.

ECONOMIC REFORMS PROCESS ON INDIAN AGRICULTURAL SECTOR

Agricultural sector is the mainstay of the rural Indian economy around which socioeconomic privileges and deprivations revolve, and any change in its structure is likely to have a corresponding impact on the existing pattern of social equality. No strategy of economic reform can succeed without sustained and broad based agricultural development, which is critical for

- Raising living standards,
- Alleviating poverty,
- Assuring food security,
- Generating buoyant market for expansion of industry and services, and
- Making substantial contribution to the national economic growth.

Studies also show that the economic liberalization and reforms prccess have impacted on agricultural and rural sectors very much.

The three sectors of economy in India, the tertiary sector has diversified the fastest, the secondary sector the second fastest, while the primary sector, taken as whole, has scarcely diversified at all. Since agriculture continues to be a tradable sector, this economic liberalization and reform policy has far reaching effects on (I) agricultural exports and imports, (ii) investment in new technologies and on rural infrastructure (iii) patterns of agricultural growth, (iv) agriculture income and employment, (v) agricultural prices and (vi) food security.

Reduction in Commercial Bank credit to agriculture, in lieu of this reforms process and recommendations of Khusrao Committee and Narasingham Committee, might lead to a fall in farm investment and impaired agricultural growth.

Infrastructure development requires public expenditure which is getting affected due to the new policies of fiscal compression. Liberalization of agriculture and open market operations will enhance competition in "resource use" and "marketing of agricultural production", which will force the small and marginal farmers (who constitute 76.3% of total farmers) to resort to "distress sale" and seek for off-farm employment for supplementing income.

MARGINALISATION OF SMALL FARMERS

A central issue in Agricultural Development is the necessity to increase productivity, employment, and income of poor segments of the agricultural population. Among the rural poor, the small farmers constitute a sizeable portion in the developing countries. Studies by FAO have shown that small farms constitute between 60-70% of total farms in developing countries and contribute around 30-35% to total agricultural output.

Liberalisation era (1990-91) began in India when over 40% of rural households were landless or near landless, and over 96% of the owned holdings and 68.53% (over 2/3rd) of owned land belonged to the size groups (marginal, small and semi-medium).

The decade of 1981-82 to 1991-92 seems to have witnessed a marked intensification of the marginalisation process-the percentage of small owners increased from 14.70% to 21.75%.

Small farmers emerged as the size group with the largest share of 33.97% in the total land, which is just doubled during this decade. As regards the Large Farmers, they were 1 % of the total owners in 1990-91 but owned nearly 13.83% of the total land. An interesting, but speculative, inference is that the changing position of the large owners represents the other side of the marginalisation process, *i.e.*, the presence, and possibly growing strength, of a small but dominant and influential group in agriculture.

Analytical reports reveal that marginalisation process could gather further momentum in the years ahead to become an explosive source of economic and political turbulence, due to the features of prevailing policy-cum-market environment in the country. Trend towards a greater casualisation (erratic and low-paid work) of the workforce that was witnessed in the 1980s appears to have continued in the 1990s. Low productivity and inability to absorb the growing labour force make the agricultural sector in India witness to a pervasive process of marginalisation of rural people. This process is likely to get intensified in the coming years, raising formidable problems in achieving sustained development of rural areas and rural people.

Both Information Technology, Genetic Engineering and Bio-Technology, which are the "drivers" of globalization with their complementarities of liberalisation, privatisation and tighter Intellectual Properties Rights, are bound to create new risks of marginalisation and vulnerability.

Information Technology is able to produce a penetrating and clinical mapping of the land, encompassing the physical, chemical and biological features, and groundwater resources, and forecast of climatic conditions in a focused manner, that even small geographical segments-the small farms-can be benefited through the guidance provided by the ways in which natural and human resources can be optimally combined with appropriate technologies, inputs and options to enhance and diversify agricultural production [KVS2K].

Information Technology will facilitate dissemination of information on development, education, extension, husbandry, marketing, production, and research, to agricultural farmers.

INDIAN AGRICULTURAL SECTOR

The Indian Agricultural sector provides employment to about 65% of the labour force, accounts for 27% of GDP, contributes 21% of total exports, and raw materials to several industries. The Livestock sector contributes an estimated 8.4 % to the country GDP and 35.85 % of the agricultural output. India is the seventh largest producer of fish in the world and ranks second in the production of inland fish. Fish production has increased from 0.75 million tons in 1950-51 to 5.14 million tons in 1996-97, a cumulative growth rate of 4.2% per annum, which has been the fastest of any item in the food sector, except potatoes, eggs and poultry meat. The future growth in agriculture must come from [GBSingh2K] *viz.*,

- New technologies which are not only "cost effective" but also "in conformity" with natural climatic regime of the country;
- Technologies relevant to rain-fed areas specifically;
- Continued genetic improvements for better seeds and yields;
- Data improvements for better research, better results, and sustainable planning;
- Bridging the gap between knowledge and practice; and
- Judicious land use resource surveys, efficient management practices and sustainable use of natural resources.

IX PLAN STRATEGY ON AGRICULTURAL DEVELOPMENT

The agricultural development strategy for the Ninth Five Year Plan is essentially based on the policy on food security announced by the Government, to double the food production and make India hunger free in ten years. The Strategy to ensure food security is as follows:-

- Doubling food production
- Increase in employment & incomes
- Supplementary/sustained employment and creation of rural infrastructure through Poverty Alleviation Programmes (PAP)
- Distribution of food grains to the people Below Poverty Line (BPL)

The Ninth Plan Target is to achieve a growth rate of about 4.5% per annum agricultural output and production of 234 MT of food grains by 2001-02. The Policy thrust and key elements of Growth strategy, as proposed in the Ninth Five Year Plan Document (Volume II: PP444), are as follows:-

- Conservation of land, water, and biological resources
- Rural infrastructure development
- Development of rainfed agriculture

- Development of minor irrigation
- Timely and adequate availability of inputs
- Increasing flow of credit
- Enhancing public sector investment
- Enhanced support for research
- Effective transfer of technology
- Support for marketing infrastructure
- Export promotion

The Ninth Five Year Plan Document (1997-2002) reveals that development of the vast rain-fed areas of about 90MH would require over Rs.37,000 Crores. Further, scientific treatment for soil and water conservation for 12 MH of arable and 3 MH of non-arable land would require about Rs.7500 Crores. Development of rain-fed areas require a substantial public investment, which may not be possible due to the new policies of fiscal compression. In the coming millennium, on the basis of current trends in the consumption pattern, the estimated total requirement of food grains is likely to be around 245 Million Tons by 2006-07.

AGRICULTURAL PLANNING AND DEVELOPMENT

India is a vast country with a variety of landforms, climate, geology, physiography, and vegetation India is endowed with regional diversities for its uneven "economic and agricultural" development, on account of (i) Agro-climatic environments (15 Zones/127 regions), (ii) Agro-ecological regions (20) and 60 sub-regions, (iii) Agro-Edephic regions, (iv) Terrain mapping sub-units, (v) Natural resources endowments (geology, geomorphology, soil, ground water, surface water, & infrastructure), (vi) Human resources (Population density), (vii) Level of investments in rural infrastructure, and (viii) Level of investment in technology and its adoption.

India has a total geographical area (TGA) of 329 Million Hectares (MH) out of which, about 265 MH represent varying degrees of potential for biological production. [Dhuruva89] report reveals that more than 50% of TGA is threatened by various types of land degradation, such as soil erosion, gully & ravine formation, salinity, water logging, shifting cultivation, etc. Development of irrigation potential is considered as the key factor in the sustenance of "Green Revolution". Despite 50 years of development planning, rainfed agriculture is the largest and the most important sector of crop production in India.

Soil resources are the most precious non-renewable vital resources for growing food, fibre, and fuel wood to meet the human needs. Management of Soil Resources is essential for both the continued agricultural productivity and protection of environment. By considering various factors like population growth rate, diminishing per capita of land and water resources, and increasing land degradation problems, it is estimated that India will be required to produce an additional 5-6 million tons of food grains annually in 21st Century. This will lead to tremendous pressure on soil resources along with competitive demand for it from industrialization and urbanization. However the capacity of soil to produce is limited and its limits to production are set by its inherent characteristics, agro-ecological settings, and its use and management.

Forests are an important natural resources of India, having a moderating influence against floods and also protecting the soil against erosion. About 95% of the forests in India is owned by States and the total area under forests is about 22% of the total geographical area.

Development of livestock has been envisaged as an integral part of sound system of diversified agriculture. In animal production, the major aim is for raising ecologically adapted animals and efficient utilization of locally available feed resource. Dairy development is intimately linked with cattle population, breed improvement, cattle health and disease management, and fodder development, etc. Animal Husbandry in India is essentially a endeavour of millions of small holders (Resource-Poor-Farmers) who rear animals on "crop residues" and "common property resources" without generally allowing them to compete with man for food grains.

The small holders produces milk, meat, wool, etc., for the community, with virtually no capital, resource, training and at a cost that no modern technology in the world had ever produced. Food and Fodder Resources will be crucial to the future development of "livestock resources" in the Country. There is very little scope for increasing the area under fodder production, keeping in view the priority for food grains, pulses and oil seeds. Development of Fodder Resources is basically an activity based on a multi-disciplinary approach involving the areas of agriculture, animal husbandry, environment & forests, revenue, rural development, and wasteland development.

Water Resources of India contain diverse group of flora and fauna. Agriculture is the greatest user of Water accounting for about 80% of all consumption. Animal Husbandry and Fisheries require abundant water. Development of Water Resources, since Independence, has been undertaken for specific purposes like irrigation, flood control, hydro-power generation, drinking water supply, industrial and various miscellaneous uses. Minor irrigation projects have both surface and ground water as their source, while major and medium projects mostly exploit surface water resources. The break up of the ultimate irrigation potential under the above three categories is,

- 58 M.Ha by major and medium irrigation projects,
- 17 M.Ha by minor surface water schemes, and
- 64 M.Ha by minor ground water schemes.

Fisheries Resources of India are either inland or marine. The principal rivers and the tributaries, canals, ponds, lakes, reservoirs comprise inland fisheries. The river extend about 27,200 kms, and other subsidiary water channel comprise about 112,000 kms. Marine resources comprises of about 2 Million sq.kms of EEZ for deep sea fishing, and 7,250 kms of coastline. With the diverse fish fauna, the development objectives are to judiciously & optimally utilize the resources for [NBFGR2K]:-

- Enhancing production and productivity of fishermen, fish farmers and fishing industry;
- Increasing fish production and thereby, raising nutritional standard of people;
- Earning of foreign exchange from export of marine products;
- Improving Socioeconomic conditions of traditional fishermen;
- Generating employment for coastal and rural poor; and
- Conservation of depleting species of fish.

Good infrastructure helps in raising productivity and lowering the unit cost in the production activities of the economy. "Agricultural Infrastructure" refers to "Rural Infrastructure" whereas "Industrial Infrastructure" refers to "Urban Infrastructure". Agricultural development requires (i) agricultural research and extension, (ii) rural financial institution, (iii) irrigation and drainage, (iv) agricultural inputs (fertilizers, seeds, credits), and (v) marketing and storage facilities.

Agriculture Credit is a crucial input for increasing agricultural production and productivity. Institutional finance for Agricultural credit is disbursed mainly by Commercial banks, Regional Rural Banks, Land Development Banks, and Cooperative banks. Share of commercial banks in total institutional credit to agriculture is about 48%, that of Cooperative banks is about 46%, and Regional Rural Banks account for 6% only. Short-term Credit accounts for 2/3rd of the total institutional lending to the Agriculture.

RAINFED AGRICULTURAL DEVELOPMENT

The overall goal of this stage is to review the important issues in rainfed agricultural development and report on the progress made in India to date. This will serve as a precursor to a detailed study to be carried out by the Indian Council of Agricultural Research, IFPRI, ICRISAT and the World Bank. That study will result in recommendations for designing a strategy to develop rainfed agriculture in India. We compare the past performance of rainfed and irrigated agriculture and of different types of rainfed agriculture, including relatively high- and low-potential areas.

We attempt to identify the factors that determine differences in performance, and we examine the possibilities for influencing those factors. Where information is not available, we recommend further analysis that may be required as a prerequisite to formulating a thorough strategy for rainfed agricultural development. We approach the problem by reviewing the relevant literature on the subject and conducting a statistical analysis of all-India district-level data. The database contains several agroclimatic and socioeconomic variables that we hypothesise to influence agricultural performance at the district level. The district-level approach, of course, does not permit analysis of the implications of micro-level diversity of rainfed agricultural systems. Therefore we focus on broader indicators with implications for area-wide development efforts. In addition to the district level data analysis, we review the literature on rainfed agricultural performance in terms of technology adoption and performance, yield levels and their variability, natural resource sustainability, and poverty alleviation.

We examine the role of economic and social policies, area development programmes and infrastructural investments in promoting sustainable rainfed agricultural development. On some topics sufficient evidence is available to draw conclusions and make policy recommendations, and on others, additional analysis is recommended.

CHARACTERISTICS OF RAINFED AGRICULTURE

Rainfed and irrigated agriculture coexist in practically every village in India. Public investment programmes, however, usually cannot be targeted so precisely. For practical purposes, they need to be planned and implemented on a larger scale, such as at the village, taluk, district or state level. For example, public programmes that provide credit, employ people, or build roads cannot target their efforts to either rainfed of irrigated agriculture; they can only target areas that are relatively more irrigated or more rainfed. Price policies, on the other hand, can attempt to target rainfed or irrigated agriculture within a given location by targeting crops that may be more likely to be rainfed or irrigated.

But few crops are either 100 per cent irrigated or 100 per cent dryland, so some spillover will always remain. Also, every crop is grown over a large geographic area, so it is difficult to isolate the socioeconomic and agroclimatic variables affecting their performance. And while price policies are important, they are not the only approach through which policy makers can influence agricultural development.

DEFINING RAINFED AGRICULTURE

In the present investigation we use the district as the unit of analysis. We do so for two main reasons. First, through a district level focus our analysis may be relevant for public investment programmes that provide infrastructure or other social services to particular areas. Second, a district focus enables us to examine the contributions of both socioeconomic and agroclimatic variables on a nation-wide basis.

The district is the smallest administrative unit for which the required data are available. To arrive at a district-level definition of rainfed agriculture, we consider the percentage of each district that is irrigated or rainfed, and consider predominantly rainfed districts as "rainfed" and pre-dominantly irrigated districts as "irrigated." Obviously there is a certain degree of arbitrariness to any threshold we may choose to distinguish between irrigated and rainfed districts.

Several previous studies have faced this same dilemma in categorising rainfed areas. Some of them have distinguished between irrigated and rainfed districts according to certain criteria such as the amount of rainfall and the level of irrigation. Some of these studies and their definitions of rainfed areas. All of these definitions suffer from the inability to distinguish between rainfed and irrigated agriculture within districts, but we accept this as an inevitable limitation. Another problem is that both the rainfall and irrigation thresholds are defined somewhat arbitrarily. We discuss rainfall and irrigation thresholds in turn.

RAINFALL CRITERIA

Bapna et al, (1984) and subsequently Jodha (1985) used broad rainfall thresholds, which is important in order not to be too exclusive. At the same time, the 500-1500 mm range maintains a degree of homogeneity in the types of agriculture under analysis by excluding both very dry, desert areas and very high rainfall areas. Such areas may face unique constraints that limit comparability to agriculture under the more moderate conditions that predominate in most of the country.

Shah and Sah (1993) and Thorat (1993) use narrow rainfall thresholds with a relatively low maximum because they intended to focus on very dry (but not quite desert) areas. It is important to note that the impact of the level of rainfall on crop production is conditioned by both the distribution of rainfall over the course of the season and the factors that determine moisture retention in a given location.1 As a result, narrow rainfall thresholds such as those used by Thorat (1993) are likely to combine some areas with disparate moisture regimes and separate others with similar moisture regimes.

A narrow rainfall range probably makes sense only if limited to relatively uniform soil types. Incorporating soil types into the definition, however, introduces yet another variable and makes the definition somewhat clumsy.1 For this reason, our preference is to utilise a relatively broad range of rainfall levels such as used by Bapna et al, (1984), excluding only those that are either desert environments or extremely humid. We choose a range of 450-1600 mm average annual rainfall as the range for predominantly rainfed districts.

The lower bound of 450 mm excludes the desert districts of western Rajasthan as well as one district each in Punjab, Haryana and Gujarat. The 1600 mm upper bound excludes the Himalayas, the northeastern states, Kerala, all the coastal districts of Karnataka and Maharashtra, and one coastal district each in Tamil Nadu and Gujarat. Where more disaggregated analysis is required to examine the performance of relatively moist or dry rainfed areas, we can subdivide the rainfall criteria into a low rainfall area (<750 mm per annum), a medium rainfall area (750-1125 mm), and a high rainfall area (>1125 mm), sometimes described as the arid, semi-arid and humid areas. To repeat the earlier caveat, these broad rainfall classes are heavily conditioned by the factors that determine moisture retention, particularly soil type.

IRRIGATED AREA CRITERIA

Classifying districts by irrigated area is difficult for several reasons. First, any threshold percentage area irrigated must be defined somewhat arbitrarily, and second, in most districts irrigated area has increased steadily during the period under study. As a result, we consider some alternate approaches to categorising districts by irrigated area. A single irrigated area threshold to distinguish between irrigated and rainfed districts.

In these studies the threshold ranges from 10 percent to 30 percent, with most defining rainfed districts as those with less than 25 percent irrigated area. 25 percent is the mean irrigated area for the years 1956-90, so according to this definition, rainfed areas are those with less than average area irrigated, while irrigated areas are those with more than average area irrigated. A number of arguments can be made about whether the figure of 25 percent is appropriate, but ultimately any definition based on such a threshold suffers from the problem that slight differences in irrigation levels will move some districts from one category to the other.

One approach is to use three categories of irrigation instead of two in order to more clearly identify the characteristics of lightly and heavily irrigated districts. This approach however, also has weaknesses, because more categories means more thresholds, which means that fewer districts will remain in one category over the entire period. As a result, when the three irrigation categories, nearly half of all districts shift categories at some point in the period.

When only two irrigation categories are used, on the other hand, only about one quarter of the districts shift categories. For that reason, we also conduct the analysis with only two irrigation categories, with districts with less than 25 percent area irrigated considered unirrigated, and districts with 25 percent or more area irrigated considered irrigated. We conduct the analysis twice; in one case all districts are analysed, and in the other only districts that remain in one category or the other throughout the study period are used.

Before continuing, we briefly discuss two other criteria for defining rainfed areas that we considered but rejected. First, districts could also be subdivided by the extent of different types of irrigation, in particular, canals, wells, or tanks. The justification for this concerns the quality of irrigation services delivered to each farm. Well irrigation, for example, is controlled by the individual farmer (to the extent that the aquifer yields water), whereas under canal irrigation farmers depend more on the amount of water taken by their upstream neighbours, so they incur a greater risk of drought. As a result, farmers with well irrigation apply more inputs and have much higher yields on average. The district-level data cover gross cropped area as opposed to net cropped area, so we control for variations in the quantity of irrigation water delivered to the farm. However, the data do not control for variations in quality or differences in farmers' response to the different levels of risk under each irrigation source. When the circumstances warrant it we may examine irrigation by source, but mainly we do not, since our main focus is on rainfed agriculture, not distinctions in irrigated types.

A second alternative criterion for defining irrigation levels would be to distinguish districts by the proportion of farmers who have access to some irrigation. In many areas, water markets or shared irrigation wells enable farmers to gain access to irrigation even if they do not own a well or are not directly serviced by a canal or a tank. As a result, the number of farmers with access to irrigation can be much larger than the number who own wells. The distinction between the proportion of area irrigated and the proportion of farmers is important because it can affect the way in which most farmers manage their crops. If a larger proportion of farmers have access to a small amount of irrigation, they may concentrate their managerial and other inputs on irrigated plots, which they may perceive to be more productive and less prone to risk of crop failure than dry plots. In this case, perhaps dryland crops would be less productive (though more farmers will be better off).

Farmers without access to irrigation, on the other hand, may devote relatively more resources to rainfed crops than those with some irrigation. Unfortunately we do not have access to district-level data on the proportion of farmers with access to irrigation, and we do not know if the proportion of farmers with access to irrigation and the proportion of area irrigated vary independently of each other across districts. Therefore we cannot consider using this definition; we raise it only to draw attention to some of the issues to consider in developing a definition of a rainfed district.

DEFINITION OF RAINFED AGRICULTURE AND ITS PERFORMANCE

The definition of rainfed agriculture present will assist us in comparing the performance of predominantly rainfed and irrigated districts. We also wish to compare the performance within rainfed agricultural districts and relate differences to a range of constraints to agricultural development.

This in turn will be useful for prioritising and organising agricultural research, public investment, and policy and institutional reform. To characterise districts according to the various agroclimatic and socioeconomic variables that constrain agricultural development, as suggested, need to construct a typology based on those variables. The typologies of Indian rainfed agriculture that already exist, the reasons why a new typology needs to be constructed, and the ways in which such a typology would be used.

EXISTING TYPOLOGIES

Current typologies of Indian agriculture are based on agroecological zones. Recently, ICAR delineated 20 agroclimatic regions based on soils and climate. 13 of these zones cover the area of this study; of the rest, 5 are in the Northeast, the Himalayas, and the Andaman, Nicobar and Lakshadweep Islands; one zone covers the high rainfall areas of the Western Ghats and the Arabian Sea coast; and one zone covers the desert in the western parts of Rajasthan and Gujarat. ICAR's agroecological zoning system is based on variations in rainfall, soil type and temperature.

Irrigation status, however, is conspicuously absent. This is a severe limitation due to the primary importance of irrigation in determining cropping patterns and productivity in most of the country.

In this analysis we overcome that shortcoming by dividing zones into primarily rainfed and primarily irrigated districts. The 20-zone system is sufficiently disaggregated to enable it to keep problems of within-zone variation to a manageable level, and the number of zones remains small enough to be manageable for most uses.

Also, the zones can be easily reaggregated for particular purposes. Recently ICAR subdivided the 20-zone typology into a total of about 50 subzones; such a disaggregated typology may be useful for certain agricultural research purposes, but for policy analysis it is too large to be functional.

In our analysis, using rainfed and irrigated districts in the 13 zones covered in our data would yield a total of 26 categories, which becomes unmanageable. For the purposes of our district-level analysis, we modify the 20-zone system so that it is sufficiently aggregated for our purposes.

We create a new 5-zone system in which each zone is a combination of two or more of the zones of the 20-zone system. The new zones, require some explanation. First, the 20-zone system does not follow district boundaries, but in our analysis districts must remain intact.

As a result, in many cases we classify a district as lying in one zone even though part of it may actually lie in another. Districts in ICAR's Zone 18 along the Bay of Bengal coast, for example, also lie in adjoining agroclimatic zones 7,8 and 12.

As a result, zone 18 drops out of the sample. Second, by combining the zones it is inevitable that some new, aggregated zones will contain substantial within-zone diversity. We find that for the case of ICAR zone 4, it makes sense to place the portion of zone 4 that lies in the Gangetic plain in one new zone, and the part that lies in upland areas of Rajasthan and Gujarat in another.

WHAT A NEW TYPOLOGY SHOULD LOOK LIKE

The existing agroclimatic typologies may be adequate for a narrow set of objectives, such as locating where certain crops are likely to be produced and which regions may be prone to certain natural resource management problems. Beyond such highly specific applications, however, the agroclimatic typologies are of limited use because they are so narrowly defined. In order to be useful for designing a strategy to develop rainfed agriculture, a typology must be constructed on the basis of the whole range of factors that affect agricultural development.

These extend far beyond simple agroclimatic conditions or even irrigation status. If a district has favourable growing conditions but lacks the infrastructure needed to support productive farming, for example, it should not be surprising to find poor performance in that district despite the favourable agroclimatic conditions. As suggested, examine the determinants of agricultural performance and demonstrate that other factors in addition to agroclimatic conditions help explain the variation in performance.

In addition to agroclimatic conditions and irrigation status, numerous additional variables can be hypothesised to influence agricultural development. Physical infrastructure such as roads and electrification, for example, and social infrastructure such as banks, markets and agricultural research and extension services, can be expected to play an important role in stimulating the agricultural sector. Demographic indicators such as population density and literacy levels also may be related to agricultural performance, as may economic policies that directly or indirectly affect input or output prices. Institutional considerations also may affect performance; they include laws governing trade, property rights, prices of inputs and outputs, etc., and the quality of services provided by government agencies. Just as there are many determinants of performance of the agricultural sector, there also are many criteria for evaluating performance. Productivity growth is one that is commonly applied, but others include the levels of poverty and food security, the variability of production and income, and the degree of degradation of natural resources.

The ideal typology of Indian agriculture would characterise regions or areas according to all the factors that determine performance over a broad range of criteria. In this way it could serve as a valuable planning tool for public investment in agricultural research, infrastructural development, poverty alleviation programmes, policy and institutional reform, etc. While such a "super typology" might be unattainable, it presents an objective to work towards. We lack the resources to develop an acceptable typology, but it remains a high priority for future research intended to support Indian rainfed agricultural development.

We analyse the determinants of rainfed agricultural development using multiple regression analysis. As suggested, identify many of the agroclimatic and socioeconomic factors that contribute to performance of Indian rainfed agriculture according to a variety of criteria. As suggested, stop short of creating a typology, but as suggested, gain preliminary indications of the kinds of information that need to go into such a typology.

INDIAN AGRICULTURAL PLANNING AND DEVELOPMENT

India is a vast country with a variety of landforms, climate, geology, physiography, and vegetation India is endowed with regional diversities for its uneven "economic and agricultural" development, on account of (i) Agro-climatic environments (15 Zones/127 regions), (ii) Agro-ecological regions (20) and 60 sub-regions, (iii) Agro-Edephic regions, (iv) Terrain mapping sub-units, (v) Natural resources endowments (geology, geomorphology, soil, ground water, surface water, & infrastructure), (vi) Human resources (Population density), (vii) Level of investments in rural infrastructure, and (viii) Level of investment in technology and its adoption.

India has a total geographical area (TGA) of 329 Million Hectares (MH) out of which, about 265 MH represent varying degrees of potential for biological production. [Dhuruva89] report reveals that more than 50% of TGA is threatened by various types of land degradation, such as soil erosion, gully & ravine formation, salinity, water logging, shifting cultivation, etc. Development of irrigation potential is considered as the key factor in the sustenance of "Green Revolution". Despite 50 years of development planning, rainfed agriculture is the largest and the most important sector of crop production in India.

Soil resources are the most precious non-renewable vital resources for growing food, fibre, and fuel wood to meet the human needs. Management of Soil Resources is essential for both the continued agricultural productivity and protection of environment. By considering various factors like population growth rate, diminishing per capita of land and water resources, and increasing land degradation problems, it is estimated that India will be required to produce an additional 5-6 million tons of food grains annually in 21st Century. This will lead to tremendous pressure on soil resources along with competitive demand for it from industrialization and urbanization. However the capacity of soil to produce is limited and its limits to production are set by its inherent characteristics, agro-ecological settings, and its use and management.

Forests are an important natural resources of India, having a moderating influence against floods and also protecting the soil against erosion. About 95% of the forests in India is owned by States and the total area under forests is about 22% of the total geographical area.

Development of livestock has been envisaged as an integral part of sound system of diversified agriculture. In animal production, the major aim is for raising ecologically adapted animals and efficient utilization of locally available feed resource. Dairy development is intimately linked with cattle population, breed improvement, cattle health and disease management, and fodder development, etc. Animal Husbandry in India is essentially a endeavour of millions of small holders (Resource-Poor-Farmers) who rear animals on "crop residues" and "common property resources" without generally allowing them to compete with man for food grains.

The small holders produces milk, meat, wool, etc., for the community, with virtually no capital, resource, training and at a cost that no modern technology in the world had ever produced. Food and Fodder Resources will be crucial to the future development of "livestock resources" in the Country.

There is very little scope for increasing the area under fodder production, keeping in view the priority for food grains, pulses and oil seeds. Development of Fodder Resources is basically an activity based on a multi-disciplinary approach involving the areas of agriculture, animal husbandry, environment & forests, revenue, rural development, and wasteland development.

Water Resources of India contain diverse group of flora and fauna. Agriculture is the greatest user of Water accounting for about 80% of all consumption. Animal Husbandry and Fisheries require abundant water. Development of Water Resources, since Independence, has been undertaken for specific purposes like irrigation, flood control, hydro-power generation, drinking water supply, industrial and various miscellaneous uses.

Minor irrigation projects have both surface and ground water as their source, while major and medium projects mostly exploit surface water resources. The break up of the ultimate irrigation potential under the above three categories is,

- 58 M.Ha by major and medium irrigation projects,
- 17 M.Ha by minor surface water schemes, and
- 64 M.Ha by minor ground water schemes.

Fisheries Resources of India are either inland or marine. The principal rivers and the tributaries, canals, ponds, lakes, reservoirs comprise inland fisheries. The river extend about 27,200 kms, and other subsidiary water channel comprise about 112,000 kms. Marine resources comprises of about 2 Million sq.kms of EEZ for deep sea fishing, and 7,250 kms of coastline. With the diverse fish fauna, the development objectives are to judiciously & optimally utilize the resources for [NBFGR2K]:-

- Enhancing production and productivity of fishermen, fish farmers and fishing industry;
- Increasing fish production and thereby, raising nutritional standard of people;
- Earning of foreign exchange from export of marine products;
- Improving Socioeconomic conditions of traditional fishermen;
- Generating employment for coastal and rural poor; and
- Conservation of depleting species of fish.

Good infrastructure helps in raising productivity and lowering the unit cost in the production activities of the economy. "Agricultural Infrastructure" refers to "Rural Infrastructure" whereas "Industrial Infrastructure" refers to "Urban Infrastructure". Agricultural development requires (i) agricultural research and extension, (ii) rural financial institution, (iii) irrigation and drainage, (iv) agricultural inputs (fertilizers, seeds, credits), and (v) marketing and storage facilities.

Agriculture Credit is a crucial input for increasing agricultural production and productivity. Institutional finance for Agricultural credit is disbursed mainly by Commercial banks, Regional Rural Banks, Land Development Banks, and Cooperative banks. Share of commercial banks in total institutional credit to agriculture is about 48%, that of Cooperative banks is about 46%, and Regional Rural Banks account for 6% only. Short-term Credit accounts for 2/3rd of the total institutional lending to the Agriculture.

Drought has multiplier effect on agricultural production during the subsequent year also, due to (i) non-availability of quality seeds for sowing of crops, (ii) inadequate draught power for carrying out agricultural operations as a result of either distress sale of cattle or loss of life, (iii) reduced use of fertilizers as the investment capacity of the farmers decline, (iv) non-availability of raw materials in agro-based industries, and (v) deforestation to meet the energy needs in domestic sector as agricultural waste may not be available in required quantity.

The Central Ministry of Agriculture (MOA) is responsible for implementation and formulation of national policies and programmes to achieve agricultural growth through optimum utilization of the land resources, water, soil, plant, fisheries, & livestock resources. Government of India implements the following agricultural related Schemes (whether Watershed based or Agro-climatic region based) in the country, which deal agricultural resources information for Planning and Development:-

- Agro-climatic Regional Planning (ACRP) Project
- Agro-Ecological Mapping Project of the National Bureau of Soil Survey & Land Use Planning (NBSS&LUP)
- All India Soil and Land Use Survey (AISLUS)
- Early Warning System of Agricultural Situation in India
- Forecasting of Agricultural output using Space, Agro-meteorology and Land based observations (FASAL) Project
- Land Records Computerisation Project
- National Agricultural Research Project (NARP)
- National Agricultural Technology Project (NATP) to strengthen research-extension-farmer (r-e-f) linkage
- National Watershed Development Programme for Rain-fed Areas (NWDPRA)

- Soil and Water Conservation Programmes
- Drought Prone Area Development programme
- Desert Development Programme
- National Wastelands Development programme
- Integrated Mission on Sustainable Development (IMSD) Programme.

EMPLOYMENT OF INDIAN AGRICULTURAL SECTOR

The Indian Agricultural sector provides employment to about 65% of the labour force, accounts for 27% of GDP, contributes 21% of total exports, and raw materials to several industries. The Livestock sector contributes an estimated 8.4 % to the country GDP and 35.85 % of the agricultural output. India is the seventh largest producer of fish in the world and ranks second in the production of inland fish. Fish production has increased from 0.75 million tons in 1950-51 to 5.14 million tons in 1996-97, a cumulative growth rate of 4.2% per annum, which has been the fastest of any item in the food sector, except potatoes, eggs and poultry meat. The future growth in agriculture must come from [GBSingh2K] *viz.*,

- New technologies which are not only "cost effective" but also "in conformity" with natural climatic regime of the country;
- Technologies relevant to rain-fed areas specifically;
- Continued genetic improvements for better seeds and yields;
- Data improvements for better research, better results, and sustainable planning;
- Bridging the gap between knowledge and practice; and
- Judicious land use resource surveys, efficient management practices and sustainable use of natural resources.

IX Plan Strategy on Agricultural Development

The agricultural development strategy for the Ninth Five Year Plan is essentially based on the policy on food security announced by the Government, to double the food production and make India hunger free in ten years. The Strategy to ensure food security is as follows:-

- Doubling food production
- Increase in employment & incomes
- Supplementary/sustained employment and creation of rural infrastructure through Poverty Alleviation Programmes (PAP)
- Distribution of food grains to the people Below Poverty Line (BPL)

The Ninth Plan Target is to achieve a growth rate of about 4.5% per annum agricultural output and production of 234 MT of food grains by 2001-02. The Policy thrust and key elements of Growth strategy, as proposed in the Ninth Five Year Plan Document (Volume II: PP444), are as follows:-

- Conservation of land, water, and biological resources

- Rural infrastructure development
- Development of rainfed agriculture
- Development of minor irrigation
- Timely and adequate availability of inputs
- Increasing flow of credit
- Enhancing public sector investment
- Enhanced support for research
- Effective transfer of technology
- Support for marketing infrastructure
- Export promotion

The Ninth Five Year Plan Document (1997-2002) reveals that development of the vast rain-fed areas of about 90MH would require over Rs.37,000 Crores. Further, scientific treatment for soil and water conservation for 12 MH of arable and 3 MH of non-arable land would require about Rs.7500 Crores. Development of rain-fed areas require a substantial public investment, which may not be possible due to the new policies of fiscal compression. In the coming millennium, on the basis of current trends in the consumption pattern, the estimated total requirement of food grains is likely to be around 245 Million Tons by 2006-07.

AGRICULTURE AND ECONOMIC DEVELOPMENT

As a country develops economically, the relative importance of agriculture declines. The primary reason for this was shown by the 19th-century German statistician Ernst Engel, who discovered that as incomes increase the proportion of income spent on food declines. For example, if a family's income were to increase by 100 percent, the amount it would spend on food might increase by 60 percent; if formerly its expenditures on food had been 50 percent of its budget, after the increase they would amount to only 40 percent of its budget. It follows from this that, as incomes increase, a smaller fraction of the total resources of society is required to produce the amount of food demanded by the population.

8

Reforms in Agriculture

Policy reforms in the rural sector are both critical and sensitive in nature as well as in their impact because of heavy concentration of working people in rural areas. At present about 40% of total state expenditure in the rural sector is absorbed by direct subsidies, another 22% by anti-poverty programmes and only 38% goes to productivity enhancing expenditure, as opposed to 60% in 1981-82.

During the post-reform period, annual growth rate of agricultural subsidies in real terms has substantially declined. But, it remains as high as 3.6% of GDP per annum. In real terms, the food subsidy increased from Rs. 10.8 billion in 1989-90 to 14.3 billion in 1995-96 at 1980-81 prices, although food subsidy as a percentage of GDP remained constant at about 0.5%.

The revamped Public Distribution System (PDS) has been introduced since 1992, which is targeted to those poor and backward regions where the Employment Assurance Scheme (EAS) is implemented (Annual Report, Ministry of Rural Areas and Employment, 1996). However, Mr. Yashwant Sinha in his Budget, 2000-01 targeted middle and upper classes. From now on, income-tax payers would not get any commodities under PDS, and the others would not get certain commodities such as sugar.

The fertilizer subsidy as percentage of total agriculture GDP declined from 0.9% in 1990-91 to 0.6% in 1995-96. But in absolute real terms, it has been subjected to high annual variations. The fertilizer subsidy policy affected different fertilizer prices differently.

The prices of phosphoric and potassium fertilisers were decontrolled in 1992, but price control on low analysis nitrogenous fertilisers continued. However, in the current year (2000-01) the subsidy price of urea is fixed at Rs. 4,600 per ton. Same for MOP (Potash) is fixed at Rs. 4,260 per ton and for Di-Ammonium Phosphate (DAP) it is Rs. 8,880 per ton. This means that there would be an increase of 15% in prices of urea and MOP and 7% in the case of DAP.

Power and irrigation subsidies, which are supported by the state governments, accounted for nearly 8% respectively of the total agricultural subsidies (as of 1994-95). During the post-reform period (1990-91 to 1994-95), rural power subsidy grew at the rate of 14% per annum in real terms, while the growth rate of irrigation subsidy remained low at 1.2% per annum (based on the Budget Estimates). But irrigation subsidy was reduced mainly on account of non-wage outlays on operation and maintenance and not because of improvement in cost recovery.

Zonal restrictions on the movement of agricultural commodities have been removed since February, 1993, including the lifting of informal controls on wheat movement by private trade.

The excise duty on coffee has been removed. The role of coffee board has diminished and now there is a trend towards open marketing of coffee. Imports of all agricultural commodities other than cereals, oilseeds and edible oils and all agricultural exports (except onion) have been decannalised.

India has agreed to phase out quantitative restrictions on import of 2700 items, out of which 800 are agricultural commodities by April 2003. An agreement had already been reached with European Union and Australia to remove quantitative restrictions on imports from these countries by April, 2000 (Haque, 1997:12).

Agricultural trade reforms initiated in 1991-92 relaxed quantitative restrictions on a few minor commodities. But by 1994-95, it included rice exports and imports of most edible oils, sugar and cotton. Nevertheless, quantitative restrictions on exports of most agricultural commodities except rice continue.

The share of tradable agricultural production protected by non-tariff barriers on the import side, were reduced from about 96% before June, 1991 to 77% in July, 1996. With the liberalisation of sugar imports under 0% tariffs, sugar industry has now to compete with imports.

Similarly, the liberalisation of cotton imports at zero percent tariff has been initiated in 1994. Edible oils are now importable at 20% tariff. The tariff on import of pulses has been reduced from 10 percent to 5 percent. In January, 1997 the U.S. and several other developed countries contended that India no longer suffers from balance of payments problems and therefore, the quantitative restrictions on imports by India would have to be removed immediately.

India made an agreement with European Union and Australia to remove quantitative restrictions on many items in three phases of 3 years, 2 years and 1 year, using April, 1997 as the reference year. But the dispute with U.S. has yet to be resolved.

In the livestock sector, state controls and subsidization of dairy co-operatives continue. The Milk and Milk products Order (1992) presents competition in the dairy industry. The poultry feed manufacturing continues to be reserved for the small scale sector.

The commercialization of fishing has been initiated. Agro-industries that export 50% or more of their output, have been allowed to import their inputs duty free and to import capital equipment at concessional import duty rates w.e.f. April, 1993. In 1991, tractors, combine harvesters and rice transplanters were included in the list of products for which there is now automatic approval of foreign equity of up to 51%.

But several other agricultural implements and farm inputs such as plastic piping, sheeting etc. are reserved for production by small-scale firms.

In the area of rural credit, the reforms include:

- Reducing target group lending from 100 percent to 40 percent in the case of regional rural banks,
- Greater freedom to the banks to rationalize their branches,
- Deregulation of interest rates of rural cooperative banks,
- Permission to urban co-operative banks to lend to borrowers in contingouous rural areas and
- Relaxation of service area restrictions.

Besides, the Reserve Bank of India (RBI) and National Bank for Agriculture and Rural Development (NABARD) have initiated actions for strengthening Regional Rural Banks (RRBs). Further, the 1996-97 budget provides for doubling the paid up share capital of NABARD and establishing agricultural development financial institutions at the state level to promote investment in horticulture, floriculture and agro-processing.

AGRICULTURE REFORMS – THE WAY AHEAD

Sustainable agriculture thus sustains rural livelihoods. This in turn is directly linked to the nation's as well as the household food security. Any development alternative to ensure long-term food security therefore has to be linked to sustainable agriculture.

Let me therefore draw the outline of the sustainable farming systems that the country needs to focus on. This is the overall framework under which location-specific alterations and adaptations need to be tried.

What is needed is a fresh approach that takes the ground realities into consideration before embarking upon any policy imperatives. I am trying to make an attempt, presenting a collection of five of the important rational decisions, which would certainly initiate the revival of Indian agriculture:

Sustainable farming: Indian agriculture faces an unprecedented crisis in sustainability. Foodgrain productivity in the food bowl, comprising Punjab, Haryana, and western Uttar Pradesh, is on the decline. The green revolution areas are encountering serious bottlenecks to growth and productivity. The dryland areas (comprising nearly 70 per cent of the cultivable lands) continue to drown in misery and apathy. Excessive mining of soil nutrients and groundwater have already brought in soil sickness. Indiscriminate use of chemical pesticides has done serious harm to environment, human health and ecology. Introducing new Centrally Sponsored Schemes or contract farming to improve production in these areas is going to be counter-productive. Banking upon genetically engineered crops to take care of the second-generation environmental impacts is sure to worsen the existing crisis. Outlays earmarked for genetic engineering in agriculture also need to be diverted to sustainable agricultural practices.

Encouraging sustainable and traditional farming practices therefore is the only way ahead. Agricultural research must reorient itself to meet the new challenges resulting from the collapse of the green revolution technology. Investments and increased outlays for agricultural research that is based on external chemical inputs like fertilizer and pesticides need to be discouraged. Instead, financial allocation should be made for reviving low-input agriculture, which uses cheap and locally available technology and in turn improves production and protects environment. This has been amply demonstrated in several parts of the world. Water productivity and efficiency has to be the hallmark of agricultural research based on the local conditions.

Local Solutions: For the past three decades, more so after the introduction of the land-grant system of education, the focus is on finding global solutions to local problems in agriculture. The World Bank/IMF, the Consultative Group on International Agricultural Research (CGIAR) and now some of the major donors like DFID and GTZ have been embarking of translocating alien approaches to agricultural improvement and have thereby exacerbated the crisis on the farm front. The Indian Council for Agricultural research (ICAR) too has blindly followed the land grant system of research and education, the negative results of which are now becoming apparent. Ignoring the traditional knowledge and time-tested technologies has created a crisis on the farm front. This process must be immediately stopped, if not reversed. Given the diversity of the agro-ecological regions, sustainable agriculture needs location-specific solutions.

International agricultural research, as well as the national agricultural research systems, should re-orient the focus of farm research based on the principles of – farmer friendly, environment friendly and long-term sustainability. Instead of the 'Lab-to-Land" approach, which has done immense damage to agriculture globally, the emphasis should be on learning from the land, meaning going back to farmers and the traditional farming systems. Technology need not always be high-tech and sophisticated. It can be simple and effective. This can only be ensured if the effort is to fit the new and improved technology to farmers need rather than asking farmers to fit into the technology package developed. This can only happen if farm research is brought back to the public sector. All technology should be freely available, and should not come with any proprietary tags.

Dryland farming: Despite the former Prime Minister Mrs Indira Gandhi's emphasis on dryland farming, agricultural scientists as well as the policy makers have failed the dryland farmers. This is essentially because the entire thrust of dryland research was to bring in an external model in which the dryland farmer, who manages to survive against all odds, would fit in. No effort was made to improve the existing technology base under numerous location-technology specifications.

At the same time, drylands continue to be plagued with recurring drought engulfing vast tracts of central and north-western India. The increased emphasis on water harvesting notwithstanding, the reduced availability of water is emerging as a major social and economic crisis.

This is because much of the investment is going into a faulty technology of rainwater harvesting, called the "Ridge to valley" system, a technology imported from the United States. In addition, the cropping pattern has to be evolved keeping in mind the water availability.

At present, more the water requirement for hybrid crop varieties more is its cultivation in the water-scarce regions. This is scandalous and unless the cropping pattern is rectified no measures to protect and preserve water resources will be effective.

IMPROVED CROPS MINE WATER

High-chemical input based technology has already mined the soils and ultimately led to the lands gasping for breath, with the water-guzzling crops (hybrids and Bt cotton) sucking the groundwater acquifer dry, and with the failure of the markets to rescue the farmers from a collapse of the farming systems, the tragedy is that the human cost is entirely being borne by the farmers. Green revolution was projected to have saved the country some 58 million hectares of additional land to be brought under the plough to produce more food, whereas almost twice that land mass has been rendered degraded and ecologically devastated in varying degrees in its aftermath.

Green revolution has not only gone sour, it has collapsed. The unexplained number of huge number of suicides a testimony to the entire equation going wrong. However, the fundamental issue of destruction of sustainable livelihoods is not at all being addressed.

All these years, for instance, the dryland regions of the country, which comprise nearly 75 per cent of the total cultivable area, have increasingly come under the hybrid crop varieties. While the crop yields from the hybrid varieties was surely high, the flip side of these varieties – these varieties are water guzzlers – was very conveniently ignored. For the sake of comparison, let us take the example of rice.

Not only rice hybrids, all kind of hybrid varieties that require higher doses of water – whether it is of sorghum, maize, cotton, bajra, and vegetables are promoted in the dryland regions. In addition, agricultural scientists have misled the farmers by saying that the dryland regions were hungry for chemical fertilisers.

The harmful combination of chemical inputs with water guzzling crops have played havoc with the drylands turning the lands not only further unproductive but also barren. The water table plummeted, the impact of deficient rainfall became more pronounced forcing farmers to abandon agriculture and migrate. As if this was not enough, Bt cotton requiring more water than hybrid cotton, was knowingly promoted so as to allow the seed industry to make profits.

Investments in rainwater harvesting need to be immediately shifted to the revival of the traditional forms of water conservation – ponds and tanks. Fodder cultivation, crop planning according to the water needs and availability and the emphasis on the local breed of cattle (and improving its productivity, rather than importing exotic breeds) need to be encouraged. Dryland crops, and that include coarse cereals, pulses and oilseeds require adequate policy measures that bring shine to these forgotten grains.

Farmers in the rain-fed areas also need to be insured against drought. This can be ensured by making it mandatory for the foreign insurance companies to invest at least 40 per cent of their funds for farm insurance.

Sugar mills: Sugarcane is the biggest threat to India's food security. The unprecedented addition of new sugar mills by successive governments has created a major crisis on the agriculture front. Requiring good fertile and irrigated land for cultivation, its growth is at the cost of staple foods like wheat and rice. With the per hectare productivity of foodgrains on the decline in the frontline agricultural states, diversion of good fertile land to sugarcane is not without accompanying hiccups. What makes the switchover to sugarcane a pernicious trend is its enormous water requirement. Sugarcane, in fact, is the biggest threat to India's food security.

Since there is no shortage of sugar in the country, and with a large number of mills actually being rendered unviable over the past two decades, an immediate ban needs to be imposed on setting up any new sugar mill. All Budgetary support to the sugar industry needs to be withdrawn as it has led to a serious environmental crisis. Reduce the area under sugarcane, improve productivity, disband most of the unproductive sugar mills, and give a new lease of life to the cane areas.

Instead the focus should shift to pulses and fodder crops. Pulses are essential for country's nutritional security and fit very well into the harsh environments. Sugarcane growers in most parts of the country can easily be made to shift to pulses cultivation given the right incentive. Such a renewed emphasis will not only help farmers and consumers alike but also rejuvenate the environment and help in restoring soil health. Pulses have the inbuilt capacity to draw nitrogen from the atmosphere.

Marketing: Providing an assured and remunerative market for agricultural producers cannot be left to the market forces. The food policy imperatives of public distribution system and announcing the procurement prices before the crop season have to be further strengthened. Agri-processing too needs to be strengthened, but not at the cost of the domestic producers. Food-processing sector should be directed to use the abundant raw material available within the country. The 'rainbow' revolution that everyone talks about is actually aimed at helping the industry to exploit the farm sector. Already a number of manufacturing units, for instance, have begun to source the agricultural raw material, including oranges, grapes, popcorn, peas etc, from America and Europe.

Domestic production in these crops is going waste. Farmers have repeatedly and in different parts of the country been dumping tomatoes, potatoes and other fruits onto the streets to express their frustration at the lack of adequate marketing infrastructure. Creating a global market for farm produce is the bane of modern agriculture. The seed multinationals, the food giants, and the supermarkets, have cornered the food chain in the process thereby destroying livelihoods, local markets and also drastically reducing food choices. Such a maket strategy has resulted in the disappearance of locally produced nutritious foods as a consequence of which micro-nutrient deficiency in human populations have grown manifold. Encouraging local markets will also reduce the dependence upon long distance transportation thereby minimising global warming. It will also help in bringing back the traditional and neglected crops, and help in changing the food habits.

Farm incomes: Growing indebtedness in agriculture is forcing an increasing numbers of farmers to end their lives. This unsavory phenomenon is a manifestation of the declining farm incomes and lack of farm credit. Institutional finance and credit has almost disappeared over the years. Banks are no longer treating agriculture for priority sector lending. Rural Banks and cooperatives are deep in the red, with a majority of them eating into their own reserves. Agriculture credit has to be revived. Schemes that encourage banks to provide easy credit facilities to farmers need to be spelled out. On top of it, agriculture credit has to be extended to sustainable farming systems. So far the banks are only providing credit for technology-oriented farming systems. This has to be extended to organic agriculture, for which an Organic Bank need to be created by NABARD (like the technology credit that goes through the private Robo Bank). Crop insurance should be extended to cover the entire farm sector immediately.

Food Security – The Way Ahead

Although, India is following the WTO dictates of doing away with the food procurement system, any tinkering with what is generally regarded as the "famine-avoidance" strategy, can be catastrophic. Corrective measures are needed to reduce inefficiency in the system while at the same time making it broad-based and widespread.

Multiple Cropping: Emphasis on commodities approach during the green revolution has encouraged monocultures, loss of biodiversity, encouraged food trade in some commodites, distorted domestic markets, and disrupted the micro-nutrient availability in soil, plant, animals and for humans humans. Thrust on farm commodities have also pushed in trade activities, encouraged food miles, adding to greenhouse emissions, water mining, and destruction of farm incomes. The need is to revert back to the time-tested farming systems that relied on mixed cropping and its integration with farm animals, thereby meeting the household and community nutrition needs from the available farm holdings.

Reverting back to mmultiple cropping will also provide the answer to the acute malnutrition that prevails in the countryside. The availability of nutritious crops, vegetables and fruits was once a part of the cropping pattern, abandoned in the wake of green revolution. The second green revolution that is being talked about will further exacerbate malnutrition crisis. This can only happen when the focus shifts away from encouraging cash crops.

For the past two decade at least, the World Bank/IMF and some other academicia and donors have been pressing developing countries to diversify from staple foods to cash crops in what is being projected as the right approach to added to farm incomes. This is a politically motivated advise and runs counter to the sustainable approached spelled out above. Many Latin American countries are faced with a serious land degradation crisis as a result. It also pushes farmers into a death trap since the developing countries do not have the resources to prôvide for adequate marketing infrastructure.

Public Distribution System (PDS) also needs to be strengthened and extended to upcoming agricultural areas in Bihar, Orissa, West Bengal and the northeast. Similarly, financial allocation must be made for assured food procurement at remunerative prices. In addition, procurement needs to be extended to coarse cereals, pulses and oilseeds to provide farmers an incentive to produce more. Food procurement operations, linked to the announcement of assured prices for agricultural commodities, were the two planks of the 'famine-avoidance' strategy that India had adopted in the wake of the green revolution. Whether the economists like it or not, the fact remains that a combination of these policies helped India to emerge from the dark days of 'ship-to-mouth' existence.

The emphasis by IMF, the World Bank and WTO to force India dismantle PDS is based on the corporate need. India's massive food procurement operations are coming in the way of the expansion of the food trade that the United States and the European Union are looking for. If the US doesn't find an assured food market in a country as huge as India, with one sixth of the world's population, the chances are that its own agriculture will collapse under the artificial weight of its own federal subsidies. That the threat is real, is clearly evident. Take a look at the recent developments in neighbouring Pakistan.

Under pressure from IMF and the World Bank, Pakistan's military government has begun lifting its decades-long support price system for key commodities – despite protests that this would be disastrous for small farmers. In India too, economists are asking the government to 'decentralise' the food procurement system, a euphemism for dismantling the PDS.

Once the government withdraws from announcing procurement prices for agricultural commodities, it is under no obligation to purchase the surplus that flows into the *mandis*. Farmers would thus be left at the mercy of the trade and the market forces, and if the past experience is any indication it simply means rendering the farming community vulnerable to exploitation thereby threatening the country's food self-sufficiency, so assiduously built over the past three decades.

The biggest crisis afflicting the farm sector is the inability to manage the agricultural surpluses. It is here that the policy planning effort has to be redirected with an effort to ensure that the surplus does not become a national liability. The approach has to be different for the rural and urban areas. Since this chapter focuses on the link between agriculture, food security and hunger, a framework for rural India is hereby proposed.

Community Grain Banks: The answer to the intricately complex, economically unsound and politically sensitive issue of public distribution rests with the poorest of the poor and is a tribute to human ingenuity, cooperation and traditional knowledge. Effectively targetting the public distribution system to reach the needy and the poorest of the poor has been a serious concern. Moreover, for several years now, the exclusion of the well-to-do beneficiaries, including income tax payers, from the provisions of the PDS have been resisted by all political parties, irrespective of their ideological leanings.

While the debate goes on, Bolangir in Orissa and Kodagu in Karnataka have demonstrated that the real beneficiaries, the poor in the villages, are not dependent upon food doles. Such a system of sharing the benefits of the harvest with the village community also exist in several other parts of the country. This is perhaps the only viable path for the nation to wriggle out of the growing threat from food insecurity.

Starvation and hunger no longer stalks a cluster of 20 villages, about 150 kms away from Bolangir town. At a time when recurring drought has brought acute misery and suffering for tens of thousand people in the district, and with the latest controversy shrouding the starvation deaths and sale of children from western Orissa showing no signs of healing, hundreds of families in and around Sundhi munda village have built a food insurance system that keeps sure hunger and death at bay. That the food security system has successfully withstood varying degrees of natural calamities and has, in fact, grown and multiplied clearly demonstrates its social relevance and effectiveness. It all began in 1990-91, when a social activist Bansi Dhar Behera, coordinator of the *Anchalika Jana*

Sewa Anusthan in Sundhi munda village, was looking for a permanent solution to mitigate human suffering arising from the non-availability of foodgrains, especially at times of distress. His appeal to fellow villagers to donate surplus paddy and rice after the harvest so as to build a grain reserve brought in 22 quintals of paddy. In all, 150 families from eight villages, almost all of them marginal farmers, responded to his call. The village grain bank was thus formed.

The grain bank became a pivot of food security. Farmers have since then deposited their 'surplus' produce with the bank after each paddy harvest. They withdraw an equal quantity of paddy at the time of need without having to pay any interest. For others, who are landless or do not have any 'surplus' for the grain bank, borrowing paddy at the time of distress is a routine. But at the time of harvest, the grains borrowed have to be returned with half a bucket of paddy as interest. For those, who cannot repay the foodgrain loan, the village *samaj* decides whether the loan can be waived or not. For the villagers, the grain bank was an escape from the clutches of the money-lenders, who often gave foodgrains to the needy to be returned in double the quantity received, and that too within three months.

Sometimes, depending upon the immediate requirement of the participating villages, the beneficiaries are asked to contribute by way of human labour. In village Batharla, a community temple and a grain store house was constructed by the beneficiaries. Their wages were paid in kind from the interest (surplus grain) that builds up over the years. In Banjupadhar village, a traditional water harvesting tank was rejuvenated for which the society distributed 16 quintals of paddy as wages. The grain bank, in other words, is also being utilised for 'food for work' programmes, all depending upon the need of the village community.

In five years, the grain bank had grown in size and volume. In 1996, the society received and disbursed 220 quintals of paddy. A year later, in 1997, it got back 253 quintals. In all, the number of people donating to the grain bank had grown by almost ten times, with a thousand families depositing paddy this year. The number of beneficiaries too increased over the years reaching 1,066 families this year, in the 20 participating villages. More than the numbers what is important is to understand that these families have perfected a social model that gives them the freedom from hunger.

The ten grain banks in Kodagu district are, however, registered under the Cooperative Act. Successfully in operation for over 30 years now, these grain banks also work on the same principle. After every paddy harvest, each member brings not less than 100 kg of paddy as their contribution to the grain bank. And during the lean months of December-January, paddy can be borrowed as loan by members. The loan is normally repaid after the next harvest with an interest of 12 per cent in terms of paddy. After the harvesting season ends, the left over paddy stocks are sold in the market. Consequently, members receive dividend varying between 10 to 20 per cent of the total share capital. Such is the underlying spirit of cooperation that like in Bolangir, each member in Kodangu district also deposits about five to ten kg of paddy every year towards what is called as the 'death fund'. The basic idea being that at times of bereavement, the village community comes to the rescue of the family in mourning. It is invariably because of the strong community ties in the villages that the grains banks have succeeded. Also, because these grain banks have remained outside the gambit of government interference. Its replication, therefore, has to be through the panchayats and the grassroot NGOs or perhaps an amalgamation of both.

Village Republics: Focus on tackling the causes of poverty, hunger, the inequitable distribution of income and low human resource base with the objective of providing everyone with the opportunity to earn a sustainable livelihood. The green revolution areas are encountering serious bottlenecks to growth and productivity. Excessive mining of soil nutrients and groundwater have already brought in soil sickness. If the livelihood of the marginalised in the society (and that in the majority world is in agriculture) it must be secured by economic activities that are sustainable, that do not threaten the integrity of the environmental assets on which they depend. Food security and hunger are directly linked to the community's control over the natural resources, and also on the long-term sustainability of the resource base.

Contrary to commonly made projections and assessments, hundreds of villages in rural India have made their own effort to chart a different but equitable path to growth and human development. Deviating from the mainstream approach, these villages have put up sign board outside the village boundary warning government officials and private company executives from entering their village. The reason: these villages have become self-reliant.

A conservative estimate based on different reports shows that close to 1500 villages have imposed self-rule and have declared themselves village republics. In these villages the residents have taken control over their natural resources – namely forest, land, minerals and water sources – and have formed strong institutions to manage them. They plan, execute and resolve all affairs inside the village and government officials and programmes are accepted only after getting approval of the residents through *Gram Sabha* (village assembly consisting of all adult members). In many such villages, the forest department, the police and other officials just execute programmes and plans chalked out in village meetings. Self-reliant villages is the answer to India's multiple and complex problems of food insecurity, hunger and malnutrition.

ADVANCEMENT IN FARMING

This fact would have surprised most economists of the early 19th century, who feared that the limited supply of land in the populated areas of Europe would determine that continent's ability to feed its growing population. Their fear was based on the so-called law of diminishing returns: that under given conditions an increase in the amount of labour and capital applied to a fixed amount of land results in a less than proportional increase in the output of food. This principle is a valid one, but what the classical economists could not foresee was the extent to which the state of the arts and the methods of production would change.

Some of the changes occurred in agriculture; others occurred in other sectors of the economy but had a major effect on the supply of food. In looking back upon the history of the more developed countries, one can see that agriculture has played an important part in the process of their enrichment. For one thing, if growth is to occur, agriculture must be able to produce a surplus of food to maintain the growing non-agricultural labour force. Since food is more essential for life than are the services provided by merchants or bankers or factories, an economy cannot shift to such activities unless food is available for barter or sale in sufficient quantities to support those engaged in them. Unless food can be obtained through international trade, a country does not normally develop industrially until its farm areas can supply its towns with food in exchange for the products of their factories.

Economic growth also requires a growing labour force. In an agricultural country most of the workers needed must come from the rural population. Thus agriculture must not only supply a surplus of food for the towns, but it must also be able to produce the increased amount of food with a relatively smaller labour force. It may do so by substituting animal power for human power or by gradually introducing labour-saving machinery.

Agriculture may also be a source of the capital needed for industrial growth to the extent that it provides a surplus that may be converted into the funds needed to purchase industrial equipment or to build roads and provide public services. For these reasons a country seeking to develop its economy may be well advised to give a significant priority to agriculture. Experience in the developing countries has shown that agriculture can be made much more productive with the proper investment in irrigation systems, research, fertilizers, insecticides, and herbicides.

Fortunately, many advances in applied science do not require massive amounts of capital, although it may be necessary to expand marketing and transportation facilities so that farm output can be brought to the entire population. One difficulty in giving priority to agriculture is that most of the increase in farm output and most of the income gains are concentrated in certain regions rather than extending throughout the country. The remaining farmers are not able to produce more and actually suffer a disadvantage as farm prices decline. There is no easy answer to this problem, but developing countries need to be aware of it; economic progress is consistent with lingering backwardness, as can be seen in parts of southern Italy or in the Appalachian area of the United States.

UNDERSTANDING CHANGING WORK PATTERNS IN AGRICULTURE

The IUF and the International Land Coalition (ILC) have jointly produced an analytical report on understanding changing work patterns in agriculture in the Ugandan sugar industry with regard to:

- Full-time waged workers; and
- Temporary and/or seasonal waged workers, including self-employed farmers hired as wage labourers.

The research aims to improve understanding of the rapidly changing patterns of production, employment and work in agriculture in order to help the IUF and ILC to more clearly focus their respective work programmes, better target their resources, and give clearer indications of potential partner organisations.

The report found:

- An increase in the number of waged workers on short-term contracts of employment on the nucleus plantation;
- Increased hiring of casual waged workers by self-employed farmers, producing sugar under contract as "outgrowers" to the sugar plantation companies;
- Increased use of casual waged workers on nucleus plantations;
- Increasing casualisation of employment.
- Ongoing downsising of the permanent waged workforce on sugar company nucleus (*i.e.,* directly managed) plantations;
- Outgrower associations acting as labour contractors, hiring casual waged labour to work on the farms of its outgrower farmer members;

The combined effects of these changes for waged workers are growing job insecurity, lower rates of pay, poorer working conditions, increasing food insecurity and growing levels of poverty. Agricultural workers have a range of responsibilities, from planting, cultivating, grading, and sorting agricultural products to inspecting agricultural commodities and facilities. They may work with food crops, animals, or trees, shrubs, and plants. Depending on their jobs, they may work outdoors or indoors.

Agricultural inspectors are employed by Federal and State governments to inspect agricultural commodities, processing equipment and facilities, and fish and logging operations for compliance with laws and regulations governing health, quality, and safety. They inspect horticultural products or livestock to detect harmful disease or infestations. To assist in eradicating disease, they also inspect livestock to help determine the effectiveness of medication and feeding programmes. They may collect samples of pests, or of suspected diseased animals or materials, and send such samples to a laboratory for identification and analysis.

AGRICULTURAL PRODUCTS WORK

Graders and sorters, agricultural products work to ensure the quality of the agricultural commodities that reach the market. They grade, sort, or classify unprocessed food and other agricultural products by size, weight, colour, or condition. *Farmworkers and labourers, crop, nursery, and greenhouse* manually plant, maintain, and harvest food crops; apply pesticides, herbicides, and fertilizers to crops; and cultivate the plants used to beautify landscapes. They prepare nursery acreage or greenhouse beds for planting; water, weed, and spray trees, shrubs, and plants; cut, roll, and stack sod; stake trees; tie, wrap, and pack flowers, plants, shrubs, and trees to fill orders; and dig up or move field-grown and containerised shrubs and trees. Additional duties include planting seedlings, transplanting saplings, and watering and trimming plants.

Farmworkers, farm and ranch animals care for live farm, ranch, or aquacultural animals that may include cattle, sheep, swine, goats, horses and other equines, poultry, finfish, shellfish, and bees. They also tend to animals raised for animal products, such as meat, fur, skins, feathers, eggs, milk, and honey. Duties may include feeding, watering, herding, grasing, castrating, branding, de-beaking, weighing, catching, and loading animals. They also may maintain records on animals, examine animals to detect diseases and injuries, and assist in birth deliveries and administer medications, vaccinations, or insecticides as appropriate. Daily duties include cleaning and maintaining animal housing areas.

FARMWORKERS AND AGRICULTURAL PRODUCTION

Farmworkers, agricultural production may have a wide range of duties, some of which overlap duties of other farmworkers. They tend to livestock and poultry; plant and harvest crops; and apply pesticides, herbicides, and fertilizers to crops. These farmworkers also repair farm buildings and fences. Other duties may include operating milking machines and other dairy processing equipment, supervising seasonal help, irrigating crops, and hauling livestock products to market. Some farmworkers operate tractors, fertilizer spreaders, haybines, raking equipment, balers, combines, threshers, and other equipment used for plowing, sowing, and harvesting. They also may help with the sorting, storage, and working in post-harvest treatment of crops.

Inspectors Field Work

Working conditions vary widely. For example, some inspectors do field work, and may travel frequently. Federal food inspectors may work in highly mechanised plants or with poultry or livestock in confined areas with extremely cold temperatures and slippery floors. The duties often require working with sharp knives, moderate lifting, and walking or standing for long periods. Many inspectors work long and often irregular hours. Inspectors may find themselves in adversarial roles when the organisation or individual being inspected objects to the inspection process or its potential consequences.

Graders and sorters may work with similar products for an entire shift, or may be assigned a variety of items. They may be on their feet all day and may have to lift heavy objects, whereas others may sit during most of their shift and do little strenuous work. Some graders work in clean, air-conditioned environments, suitable for carrying out controlled tests. Some may work evenings or weekends because of the perishable nature of the products. Overtime may be required to meet production goals. For farmworkers in nurseries, work is seasonal; spring and summer are the busier times of the year and hours in the cold weather tend to be fewer.

These workers enjoy relatively comfortable working conditions while tending to plants indoors. However, during the busy seasons when landscape contractors need plants, work schedules may be more demanding, requiring weekend work. Moreover, the transition from warm weather to cold weather means that nursery workers might have to work overtime with little notice in order to move plants indoors in case of a frost. Farmworkers enjoy a somewhat inde-pendent lifestyle working with animals or on the land. Benefits include the wide-open physical expanse, the variability of day-to-day work, and the rural setting. However, hours are generally uneven and often long; work cannot be delayed when crops must be planted and harvested, or when animals must be sheltered and fed. Weekend work is common, and farmworkers may work a 6- or 7-day week during planting and harvesting seasons.

About 1 out of 5 agricultural workers had variable schedules, compared with fewer than 1 in 10 workers in all occupations combined. As much of the work is seasonal in nature, many workers also obtain other employment. Migrant farmworkers, who move from location to location as crops ripen, live an unsettled lifestyle, which can be stressful. Much farm and ranch work takes place outdoors in all kinds of weather and is physical in nature. Harvesting fruits and vegetables, for example, may requiremuch bending, stooping, and lifting. Some field workers may lack adequate sanitation facilities, and their drinking water may be limited.

The year-round nature of much livestock production work means that ranch workers must be out in the heat of summer, as well as the cold of winter. Those who work directly with animals risk being bitten or kicked. Farmworkers in crop production risk exposure to pesticides and other potentially hazardous chemicals that are sprayed on crops or plants. However, exposure is relatively minimal if safety procedures are followed. Those who work on mechanised farms must take precautions when working with tools and heavy equipment to avoid injury. Agricultural workers constitute by far the largest segment in the unorganised sector and their number according to 1991 Census was 74.6 million. In addition, a significant number, 110.7 million, are listed as cultivators (large, medium and small) of whom approximately 50 per cent belong to the category of small and marginal farmers. Many of these small and marginal farmers on account of utterly deficit, small and uneconomic holdings and low yield work on the land of others. Further, a significant number engaged in livestock, forestry, fishing, orchards and allied activities as well as small and marginal farmers work as agricultural workers in their spare time or in times of difficulty to supplement their meagre incomes.

In spite of the fact that these agricultural workers have such numerical strength, they are extremely vulnerable to exploitation on account of low levels of literacy, lack of awareness, persistent social backwardness and absence of unionisation and other forms of viable organisation. The avenues of stable and durable employment for them have been limited leading to inter-district and inter-state migration in search of better avenues of employment and wages but with a lot of dislocation of family life, dislocation of education of children and numerous other handicaps. Several measures have been taken to protect the interests of the working class and uplift the condition of agricultural workers. The very first legislation, the Minimum Wages Act, 1948 was applied to the agricultural sector also.

Subsequently, the Plantation Labour Act, 1951 was enacted to provide certain basic facilities to plantation workers. Many other existing labour laws are applicable or have direct bearing on agricultural labour. The problems of agricultural labourers have been sought to be tackled through Multi-dimensional course of action *viz.*, impro-vement of infrastructural facilities, diversification to non-farm activities, skill improvement programmes, financial assistance to promote self-employment, optimising the use of land resources, etc., through a variety of rural development, employment generation and poverty alleviation programmes.

All these efforts have not been able to adequately protect the interests of agricultural workers. This is partly on account of lack of bargaining power. Keeping in view this broad perspective, the Ministry of Labour is contemplating to bring a comprehensive legislation to safe guard the interests of agricultural workers.

Service Conditions of Agricultural Workers

The proposed legislation would provide for regulation of the service conditions of agricultural workers and provides for certain welfare measures which include financial assistance in case of death and injury, payment of group insurance premia, health, maternity benefits, old age pension, housing assistance and educational assistance to the children of agricultural workers. Special provision/welfare schemes for women workers prohibiting their employment after sunset, rest shelter with employment of 20 and above female agricultural workers for use of children under the age of six, ensuring payment of equal wages to men and women for same and similar nature of work as required under Equal Remuneration Act, maternity benefits, etc., are also provided in the proposed legislation. To meet the expenditure for various welfare measures there is provision for constitution of an Agricultural Workers' Welfare Fund at the district level to be financed by employers' contribution and contribution by the workers. The proposal is at the stage of consideration at various levels in the Government. However, Government's endeavour is to finalise the proposal at the earliest. For the benefit of the Agriculture workers Government has launched Krishi Shramik Samajik Suraksha Yojana from 1.7.2001 to provide social security to the agriculture workers.

Common Agricultural Policy

One of the principal objectives of the common agricu-ltural policy (CAP) is to provide farmers with a reasonable standard of living. Although this concept is not defined explicitly, one of the measures tracked within the policy is income development from farming activities. Economic accounts for agriculture provide information that allows an analysis of agricultural activity and the income generated by it.

This stage gives an overview of recent changes in agricultural output, gross value added and prices in the European Union (EU), and their effect on income from agricultural activity. The EU-27's agricultural industry generated EUR 125 400 million of gross value added at producer prices in 2009, which represented a 14.0 per cent reduction in relation to the previous year. There were large decreases in both the value of crop output (down 13.9 per cent to EUR 171 000 million in 2009) and animal output (down 10.9 per cent to EUR 133 000 million); these were partly compensated for by a sizeable reduction in the value of intermediate consumption of goods and services (down 10.5 per cent).

Changes in the value of agricultural output comprise a volume and price component: one important strand of recent changes in agricultural policy has been to move away from price support mechanisms, so that prices more accurately reflect market forces and changes in supply and demand. During the period 2005 to 2009 there were considerable differences between the Member States in the development of deflated agricultural output prices: such deflated prices show the extent to which agricultural prices have changed compared to consumer prices. Deflated prices rose in nine of the 26 Member States (Germany, no information available), the largest increases being recorded for the United Kingdom (average growth of 5.9 per cent per annum), Cyprus (3.9 per cent per annum between 2005 and 2008) and Romania (3.2 per cent per annum), while reductions were posted in 17 of the Member States, the most significant being in Latvia (-6.3 per cent per annum), Slovakia (-6.2 per cent per annum) and Estonia (-6.1 per cent per annum).

The development of deflated agricultural input prices showed a very different picture, as prices rose in 17 of the 25 Member States for which data are available between 2005 and 2009 (Germany and Ireland, no information). As with output prices, Cyprus (4.9 per cent per annum between 2005 and 2008) and the United Kingdom (4.7 per cent per annum) reported the highest input price increases, followed by Portugal (3.5 per cent per annum). There was an overall 9.4 per cent increase in EU output prices for agricultural products between 2005 and 2009, with a breakdown between crop output (9.0 per cent) and animal output (9.8 per cent) showing prices increasing by a similar magnitude. The overall increase in output prices between 2005 and 2009 did not occur as a stable development, as there was a considerable reduction in prices between 2008 and 2009 when the price of agricultural products fell by 13.9 per cent. The largest reductions between 2008 and 2009 were recorded for cereals (-30.2 per cent), eggs, milk, fruits and olive oil (reductions of between 14 per cent and 17 per cent).

The real net value added at factor cost of the agricultural activity per unit of labour (expressed in annual work units – equivalent to the work performed by a person employed full-time), also termed as agricultural income indicator A, declined by 11.7 per cent in the EU-27 in 2009, compared with 2008. There were stark contrasts among the Member States, with decreases in income of more than 20 per cent in Hungary, Luxembourg, Ireland, Germany and Italy, contrasting with rapidly rising incomes in Malta (7.1 per cent) and Denmark (4.2 per cent).

INDIA'S ECONOMIC REFORMS IN AGRICULTURE

Although India's economic reforms were initiated in June 1991, the process of liberalisation was implemented gradually and thus it is difficult to assess the full impact of the liberalisation measures. Nevertheless, an attempt is made to discuss what is observable in terms of agricultural growth. One observation is that the expected increase in exports due to liberalisation simply did not occur. India's share in world exports was 0.6 per cent in 1997; India has to aim for at least 4 per cent by 2005 in order to meet the growing import demands for capital goods, raw materials and crude oil as well as to meet her external financial commitments. For the last decade or so, India's share in world exports of agriculture has been between 2 per cent and 3 per cent.

India is not as competitive as the other countries and calculations show that India's crop yields have increased at a slower rate over the 1990s. In addition, the agricultural sector's output growth decreased to 2.9 per cent during 1992-1993 to 1998-1999. Kalirajan and others (2001) explain that two important reasons for the slowdown are that there was no major breakthrough in developing new high-yielding varieties during the 1990s and there was a decline in the environmental quality of land which reduced the marginal productivity of the modern inputs. What could this mean in terms of the effectiveness of the policies of reduced protection to industry, a market determined exchange rate and the opening of the agricultural sector to foreign trade? First, although the reduction to protection of industry is substantial, there is reason to believe that the reduction was not necessarily sufficient to benefit the agricultural sector whose tariffs were also drastically reduced. Hence, the expected shift in resources to agriculture did not occur. Second, is the apparent ineffectiveness of the market determined exchange rate in boosting exports.

This is however not surprising as the exchange rate may not be a key factor determining agricultural export demand for India. In general, unlike manufacturing industries, agriculture did not benefit much from these two policies because the share of imported inputs in the value of agricultural production is small. It is likely that a change in the mindset and attitude of farmers has yet to take place and there are delays or hesitation in embracing India's openness. Third, in opening up the agricultural sector to foreign trade, India has taken major steps towards trade liberalisation since 1991, partly on its own initiative and partly from its commitments to WTO. Kalirajan and others (2001) provide a detailed review of these reform procedures.

But why have the benefits from trade liberalisation been slow to come? One reason is that prospects for growth in agricultural exports depend partly on domestic policies and partly on the removal of protectionist policies pursued by developed countries such as Japan and members of the European Union (EU). An OECD report (1998) estimated that the producer equivalent subsidy in the OECD countries increased by US$ 9.3 billion from 1988 to 1993 and this subsidy as a percentage of the value of production in 1997 was 9 per cent in Australia, 20 per cent in Canada, 47 per cent in EU and 70 per cent in Japan.

These protectionist practices do not seem likely to come to an early end. An UNCTAD report (1999) noted that 29 member countries of the OECD spent an average of US$ 350 billion a year in agricultural support between 1996-98. Schumacher (2000) further reports that the EU provides product-specific trade distorting domestic support to at least 50 different agricultural products. The implication of these reports is that food exports from India may not show a large increase given the international environment and the still-existing restrictions on exports in the major importing markets based on the self-sufficiency argument and food security.

Other macroeconomic factors, such as the recession in developed countries in 1996-98 as well as the 1997 South-East Asian financial crisis, have clouded the possibilities of increasing Indian exports. Another problem faced by Indian agricultural exporters is the protectionist measures in the form of non-trade barriers that developed countries use to restrict market access. This is by tightening requirements of quality, testing and labeling, and anti-dumping and countervailing measures. For example, in May 1997, the EU banned marine products from India citing unhygienic processing conditions.

The extra costs of meeting the standards required in export markets as well as costs associated with changes in the production mix and transactions associated with exports may well be discouraging Indian exporters. One existing problem of India's agricultural protection is the use of input subsidies. The general argument favouring this has been that it is necessary to encourage the use of particular inputs for production for various benefits. For India, Gulati and Sharma (1995) show that the input subsidy in per cent of GDP increased from 2.13 in the triennium ending 1982-1983 to 2.73 in the triennium ending 1992-1993.

But the benefits of these subsidies have accrued to only certain classes of farmers in some regions cultivating irrigated crops. Furthermore, highly subsidised prices of inputs such as irrigation water and electricity for pump sets have encouraged cultivation of water-intensive crops, over-use of water, ground water depletion/salinity and water logging in many areas. Subsidy for nitrogen fertilizer on the other hand has resulted in nitrogen phosphorous potassium imbalance and acted as a disincentive for use of the environmentally friendly organic manure. As a result, the linkage between food crops and non-food crops, which include fodder, has been reduced. These adverse consequences are a drain on the fiscal burden of central and state Gove-rnments.Thus, if not properly monitored, input subsidies can be counterproductive and, in this context, protection to lower costs of production should be done selectively in the course of liberalisation.

In fact, Agenda 21 of the United Nations Conference on Environment and Development in 1992 stressed that there is a need for integration of environmental considerations in the pricing of natural and other resources in such a way that prices reflect social costs. Such a pricing policy will not only lead to a more efficient use of scarce resources but also result in subsidy reductions and improvements in environmental quality. The money saved from the reduction of subsidies can be spent in the development of rural infrastructures, agricultural research, farmers' education and other forms of support for agriculture.

AGRICULTURAL GROWTH AND PERFORMANCE

While the above analysis has provided a general view of the impact of economic reforms, this chaper examines agricultural growth and performance in the states of Bihar, Karnataka, Tamil Nadu and Punjab with their attendant policy implications. The yields for various crops in these states differ greatly. While Tamil Nadu had the highest yield in rice, oil seeds and sugarcane, Punjab enjoyed the highest yields in wheat, coarse cereals, pulses and food grains. Karnataka on the other hand is seen to do well in cotton and Bihar performed quite well in pulses and coarse cereals. Further analysis and findings by Kalirajan and others (2001) show that Punjab had made remarkable achievements on the agricultural front while Bihar had remained stagnant in the last two decades, with Karnataka and Tamil Nadu showing moderate achievement.

Clearly, differences in physical endowments, climatic conditions and institutional characteristics are some of the reasons for the varying productivity performance. Thus, having across the board economic reforms is likely to work less effectively than state-specific policy measures that enable each state's agricultural yields to reach their full potential.

The comparative advantage of each state's agricultural production should be determined and with inter-state restrictions removed, total agricultural output would see a very significant increase. For example, Karnataka with less favourable soil and water resources should be given incentives to concentrate on agro-processed products and corporate agriculture in horticulture, floriculture and animal husbandry, or to undertake watershed development to help with dry land agriculture. Many studies have indicated that with watershed areas, productivity growth has been mainly due to seed and fertilizer use.

Thus, this state has to be given input subsidies for high yielding seed varieties but at the same time, the farmers need to be educated on the over use of chemical fertilizers. With Bihar, agricultural performance is problematic on many fronts. First, although demographic pressure has increased and agricultural technology has improved, most of the uncultivated land is concentrated in southern Bihar, where irrigation facilities have not kept pace and the soil is of poor quality. Given the physiography of southern Bihar, wells are also unsuitable and thus the dominant mode of irrigation has been through tanks whose expansion and maintenance has been neglected.

Second, the infrastructural facilities of Bihar have been lagging as seen by the infrastructure development. Due to infrastructural bottlenecks, availability of modern goods and services has not increased or their supply remains costly or unreliable. Third, agriculture in Bihar is dominated by small and marginal farmers and the prevalence of mass poverty is largely related to the backwardness of agriculture. Fourth and importantly, the state agricultural policies in Bihar are in dire need of review. The semi-feudal production condition still exists in rural areas and the ineffective protection of tenancy rights has hindered agricultural growth. The slow pace of land consolidation reflects inadequate financial outlays and a shortage of manpower. Kalirajan and others (2001) note that marketing and extension services in Bihar are also rather weak compared to the other states.

Punjab on the other hand, was one of the few states which enjoyed the success of land reforms and the high priority of investment in rural infrastructure.Also, the irrigation base of the small and medium sized farms was comparable to that of large farms. In addition, the Punjab Agricultural University at Ludhiana contributed to the development of new seed varieties. However, there are clear signs of a decline in crop yields since the 1990s and this has been associated with the increasing use of fertilizers and excessive water use which have increased the unit cost of production as a result of declining soil quality.

Hence, care is needed when providing further input subsidies in fertilizer and water use. Another related fact is the steep increase in wages in Punjab and in the absence of productivity increases, the cost increase has affected the profitability of farmers. With Tamil Nadu, the main crop has been rice as this state is blessed with two monsoons. But from 1992-1997, there has been a steady decline in the areas irrigated by canals and an increase in well-irrigated areas while the use of tanks remains an unreliable source of irrigation. However, major improvements in about 10 rice varieties released in the early 1990s can be expected to improve productivity growth in rice production although pests and diseases as well as imbalance in the use of fertilizers are major constraints.

Thus Tamil Nadu could do with subsidies of pesticides and farmers should be educated on the more effective use of fertilizers to obtain high yields. Interestingly, the cropping pattern of late has shown increasing substitution of food crops by commercial crops but there is concern that the benefits will reach farmers only with the development of adequate infrastructure such as roads and markets. However, that Tamil Nadu has a higher index than the all India average of infrastructure.

CHALLENGES OF GLOBALISATION

It is important to realise that globalisation poses many challenges to a developing country like India, which had relied on a state directed and regulated policy regime for more than four decades. In moving to a more open, market-based economy there are many transitional problems that the country has to manage. The Government must play a pro-active role in facilitating the globalisation process so that the opportunity sets for the economic agents are widened and the adverse effects of globalisation are minimized. The Indian Government must also prepare the necessary information base and develop its capacity to articulate India's concerns and policy trade-offs in the international forums for multilateral trade and environmental negotiations. In addition, the Government should embark on an extensive programme to educate farmers on the need to meet the standards required in the export markets.

In fact, India needs to seek technical assistance in creating the capacity for meeting such standards and to consider watershed developments for environmental considerations. Equally important is the need to disseminate information about possible export markets to farmers, so that market access is achieved at minimum cost. Given the requisite information about markets and profitability, the likelihood of farmers investing in post-harvest and processing technologies and storage and efficient transportation arrangements as well as developing supporting infrastructure is very high. Although the brave and bold move by India to reduce the tariff rate for agricultural products from 113 per cent in 1990-1991 to 26 per cent in 1997-1998 deserves to be applauded, the question of whether India is ready to compete in world markets remains to be seen.

The infant industry argument may still hold for India to shield itself from external competition but one can easily question the length of time that is required to that end. Also, a delay in opening up to foreign trade has the danger that local producers may become too complacent and never be ready for competition. As India opens up externally, it is also expected to face vulnerability in the wider international price fluctuations and thus Acharya (1998) claims that a minimum price support scheme is important.

These prices can also act as a signal to adopt modern inputs and invest in yield-raising infrastructure for increasing production. For instance, keeping basic staple food grains at reasonable prices would induce farmers to switch over to high value crops. However, during the 1990s, Kalirajan and others (2001) shows that procurement prices especially for rice and wheat have been increasing faster than the general price level. Such high prices along with guaranteed purchases by the Food Corporation of India have pushed up market prices.

These higher prices are partly responsible for the large buffer stocks with the Food Corporation. If this trend continues, India's comparative advantage will be eroded. With openness and high price instability, unstable export revenue can also be expected. One way of reducing such risk is for India to diversify her agricultural exports. For example, since 1990, even in commodities such as tea, coffee, cocoa and spices, where India is supposed to have a comparative advantage international prices have been unstable. Besides increasing the type of exports to obtain more export revenue, India should also seriously consider exporting more value added agricultural products through agro-processing such as processed vegetables, fruits, fish and meat products given that export or even local demand for basic agricultural products would decline as incomes rise.

The move to higher value added activities within the agricultural sector also spells greater opportunities for industrialisation and vice versa as borne by Kalirajan and Shand's (1997) findings of a bi-directional relationship between agriculture and industry for most Indian states. On the other hand, Sivakumar and others (1999) establish empirical evidence of high forward linkages of agriculture due to the presence of agro-industries while Satyasai and Viswanathan (1999) show the significance of the spillover effects to the industrial sector via the intensive use of purchased inputs in the agricultural sector. The lack or slow pace of internal or domestic liberalisation is also seen to hinder the possible gains from external or trade liberalisation. For example, although central zoning restrictions have been abolished, state government restrictions on inter-state and even inter-district restrictions on marketing and movement of goods still exist in many cases. This interferes with the benefits from crop speci-alisation and economies of scale arising from comparative advantage. The land market is another example of distortion whereby land ceilings exist preventing the operation of large-sized farms.

This has led to the emergence of a large number of small economically unviable land holdings. The easy leasing of land should be permitted with assurance of resumption. Yet another problem lies with the insufficiency of credit to agriculture. From 1995-1996, the Rural Infrastructure Development Fund was set up to allocate funds for the completion of projects and the government has committed itself to strengthening the cooperative credit structure through substantial refinancing and restructuring of the Regional Rural Banks. However, due to varying institutional factors in the Indian states, these domestic reforms can be expected to yield quite different results.

Although India missed the opportunity to open up two decades ago, its attempts to do so now must be regarded as better late than never. Others such as Desai (1999) observe that, "the logic of the global economy as well as India's interests dictate that India become proactive in its liberalisation policies. India must liberalise not because it has no choice but because it is the best choice". His lament that India has adopted a 'victim mentality' when it really needs to adopt a 'winner mentality' has become less of a concern as over time, India has shown commitment to stay on the bandwagon of globalisation. Having realised that globalisation is a necessary but not a sufficient condition for high growth production, India has undertaken economic reforms, both internal and external.

However, it must be ensured that these reforms are synchronised so that the pace of both reforms is set right in order to work hand in hand to promote agricultural productivity growth. Thus, training the farmers and educating them appropriately to change their mindset and reorienting them to take up new activities or adopt foreign technology is of utmost importance. In this context, it is necessary to involve non-governmental organisations in training and mobilising the rural poor to face the challenge of liberalisation.

Also, with domestic economic reforms, more care needs to be exercised to draw up state-specific liberalisation measures to maximize their benefits. Lastly, in the implementation of these reforms for successful globalisation, one crucial element, not entirely within control is the need for good governance and stability in the political and economic environment. Political leaders who are the ultimate decision makers in these matters need to examine their own role dispassionately. It is quite apparent that at this relatively early stage, there is little observable evidence of gains to India's agricultural performance after opening up.

However, there could easily be benefits that have not yet surfaced, or are yet to be identified and perhaps too difficult or intangible to measure. Whatever the case, it is highly likely that it is too soon to assess the full impact of globalisation and economic reforms. Furthermore, the process of liberalisation has been gradual and remains incomplete. For example, the complete removal of quantitative restrictions after March 2001 will have provided an opportunity for Indian farmers to tap world markets and, if they are successful, results should start to become evident soon.

Export promotion via the development of export and trading houses as well as effective liberalising export promotion zone schemes for agriculture are fairly recent measures and only time will tell as to how effective these measures are. Other possibilities such as agro-industry parks for promoting exports are also in the pipeline. In conclusion, India has successfully set sail on the waters of globalisation and economic reforms and even in the wake of economic and political instability, she has to carefully steer her course in order to reap the benefits of increased productivity growth in the agricultural sector.

ACCELERATING AGRICULTURAL GROWTH

Poverty in India is predominantly a rural phenomenon. About 70 per cent of the population, and about 85 per cent of the poor, live in rural areas and most depend on agriculture. Agriculture provides livelihood to 60 percent of the rural people and remains vital for food security.

KEY ISSUES AND CONSTRAINTS

Deceleration in agricultural growth. In the last decade (1995/96-2004/05) the agricultural GDP growth rate slowed to less than 2 per cent per year, compared to about 3.5 per cent per annum in the preceding decade. In the poorest states, such as Madhya Pradesh, Orissa, and Rajasthan, growth in the last decade was below 1 per cent per year. The stagnation of agriculture, the high proportion of poor dependent on it, and the widening rural-urban income gaps are major concerns of the Government of India (GOI).

Low Agricultural Productivity. With fixed land and water availability, higher agricultural growth can be achieved only by increasing productivity per unit of these resources through effective use of improved technology. Per hectare yields of major crops (foodgrains, oilseeds, other cash crops) in India are lower than many other major producing countries and well below those obtained in crop trials within India. Moreover, crop yields in many of the poorest states (Bihar, Orissa, Rajasthan, North-east states) are far below those of richer states (Punjab, Haryana, Tamil Nadu). The virtual collapse of the agricultural extension system in most states limits farmer access to improved technologies and practices. Weaknesses in the agricultural research system, such as an irrigated agriculture bias, weak prioritisation and proliferation of programmes, top-down supply driven programmes, weak cost effectiveness, undermine the effectiveness of the public agricultural research system. Natural resources degradation.

Sustainability of the land and water resource base is at risk with increasing soil degradation and overexploitation of groundwater in many areas. In addition to erosion, salinity and alkalinity, soils are also losing carbon and micronutrients due to unbalanced fertilizer use. Nearly 30 per cent of the blocks in the country are presently classified as semi-critical, critical or overexploited as groundwater use exceeds the rate of groundwater recharge. Composition of public expenditures. Public expenditures on agricultural subsidies including for power, irrigation and fertilizers, have crowded out productivity-enhancing investments, such as for irrigation development, agricultural research and extension, rural roads and electrification.

According to the Planning Commission, budgetary subsidies in agriculture increased from around 3 per cent of agriculture GDP around the late-1970s to about 7 per cent in the early 2000s. During the same period public investment in agriculture declined from 3.4 per cent of agriculture GDP to 1.9 per cent.

GOVERNMENT OF INDIA'S RESPONSE

The Government of India's (GOI) goal is to raise agriculture's growth rate to 4 per cent per year in the 11th Plan.

To achieve this, the Plan aims to:

- Double the growth rate of irrigated areas and improve water management in rainfed areas.
- Reclaim degraded land.
- Bridge the knowledge gap through effective extension.
- Foster diversification to higher value agricultural products.
- Promote animal husbandry and fisheries.
- Facilitate access to credit at affordable rates.
- Improve the incentive structure for markets.

Water Resource Management and Irrigation. Water is a critical input for achieving higher agricultural growth and ensuring greater food security. Presently about 40 per cent of cultivated area in India is irrigated. GOI is putting emphasis on more effective use of existing irrigation potential, expansion of irrigated area, and better water management in rainfed areas.

The GOI Accelerated Irrigation Benefits Programme aims to assist states to finish the construction of uncompleted major and medium irrigation schemes. Many states, however, lack the policy, regulatory, and institutional framework for efficient, sustainable, and equitable allocation and use of water, or for capturing the environmental costs of inefficient use. Many states often do not allocate sufficient public funds for operations and maintenance of canals, which lead to the rapid deterioration of irrigation canals and reduce the availability of water to farmers.

Limited cost recovery also limits funds for operations and maintenance and undercuts farmer incentives to use water more efficiently, leading to waterlogging and salinity problems in some areas. Some states have adopted participatory irrigation management on a wide scale to improve the management and sustainability of surface irrigation systems. Free power to farmers by some states or highly subsidised power encourages the excessive use of ground water. This has led to the increased proportion of over-exploited areas in the country and large fiscal costs to state governments.

The political economy of power sector reform however impedes progress in this area. Rainfed areas, where large numbers of poor live, cover 60 per cent of cultivated area in India. Watershed management, rainwater harvesting and ground water recharge can help augment water availability in rainfed areas. The GOI recently established the National Rainfed Areas Authority to develop an action plan for rainfed areas.

In moving forward, some challenges that would need to be addressed include:

- Multiplicity of programmes and agencies (MoA, MoRD, MoEF, donors) and lack of consistency among programmes within the state (scope, community partici-pation, contributions);
- Weak coordination among various agencies; and
- Weak local institutions to ensure effective implementation. Bridging the Knowledge Gap.

The GOI is providing additional resources for the National Strategic Research Fund. However, it would be critical to improve the governance and implementation structure of the Indian Council of Agricultural Research system to improve its research effectiveness. The Ministry of Agriculture (MoA) is seeking to improve extension effectiveness by scaling up the Agricultural Technology Management Agency (ATMA) approach which builds on a bottom-up approach to extension planning and imple-mentation.

It would be important to build on the factors of success derived from the ATMA pilots which include decentralised decision making and funds flow, priority to institutional capacity building and training of extension staff on the new bottom-up approach to extension planning and implementation and ensuring convergence of funds from the various departments at the district level. The GOI is also encouraging states to develop district plans for agricultural development based on agro-ecological potential. These district plans will be the basis for the formulation of the state agricultural development plans which would be supported under the proposed additional central assistance of ₹. 250 billion announced by the Prime Minister in May 2007. Ensuring adequate stakeholder consultation and community participation in the development of these plans would strongly influence the success of this programme. Diversification of Production and Income Sources. Rising incomes are driving the increased demand for higher-value fresh and processed agricultural products in domestic markets and globally. This opens new markets and income opportunities for farmers. Many of these commodities, however, are highly perishable and inadequate rural infrastructure (markets, rural roads, electrification) and services (market information, risk management) increase losses and inhibit the more rapid expansion of high value production.

The GOI recently launched the National Horticulture Mission — the single largest agriculture plan scheme — to promote increased horticulture production. The Bharat Nirman Programme, which supports the expansion of irrigation, rural roads, electrification and telecommunication) will help reduce losses and marketing cost and improve the competitiveness of local products. The MoA is also encouraging states to amend their Agricultural Produce Marketing Committee (APMC) Acts to liberalise the marketing of agriculture produce at the state level and improve farmer access to markets. So far 11 states have adopted the amendments.

In addition, MoA is also promoting public-private partnerships to develop markets and is providing investment grants to encourage investments in agro-processing and marketing. Continuing trade and domestic regulatory reforms are encouraging increased investments in rural areas. These include the removal of most domestic marketing controls—storage, movement, credit, wholesale marketing, and food processing. However, reforms of the institutional and policy framework for land are needed. Many states have computerised their land records/registration, but reforms of the institutional and data structures are essential to reduce the cost of land transfers and enhance tenure security in rural and urban areas (for example, eliminate overlapping property claims and facilitate access to credit).

The bans or restrictions on land leasing limits access to land by poor and landless rural households and drives tenancy underground. They also limit the productivity of land use. Diversification of incomes through rural non-farm activities is an important element of the overall strategy to address rural poverty.

More rapid growth of the rural non-farm sector is, however, constrained by government interventions in labour, land, and credit markets and inadequate and poor quality infrastructure and services. To improve access to credit in rural areas the GOI is supporting reforms of the credit cooperative system. Ensuring Food Security.

While pursuing crop diversification, an emerging concern is the stagnation in foodgrain production which is raising food security concerns. The 11th Plan approach paper notes that to achieve the goal of a 4 per cent overall agricultural growth rate per year, foodgrain yields per hectare need to increase by at least

2 per cent to 3 per cent per year to compensate for the possible shift in cultivated area from foodgrains to high value crops. To enable this, a concerted effort is needed to raise the relatively low foodgrain yield levels in Eastern and Central India in particular. Improving Public Expenditures. In recent years the GOI and state government have exerted greater effort to cap the amount of the various subsidies, while increasing expenditures on more productivity enhancing investments.

The latter includes several centrally sponsored programmes, such as the National Horticulture Mission, Bharat Nirman rural infrastructure programme, and the National Rural Employment Guarantee Scheme. Participation of the poorest states in these programmes is hampered, however, by lack of capacity to develop appropriate plans and implement them effectively.

THE ROLE OF GLOBALISATION AND ECONOMIC REFORM

The Indian agricultural sector has been undergoing economic reforms since the early 1990s in the move to liberalise the economy to benefit from globalisation. Analyses its effects on agricultural productivity and growth and discusses the problems and prospects for globalisation to draw policy implications for the future of Indian agriculture.

India, which is one of the largest agricultural-based economies, remained closed until the early 1990s. By 1991, there was growing awareness that the inward-looking import substitution and overvalued exchange rate policy coupled with various domestic policies pursued during the past four decades, limited entrepreneurial decision making in many areas and resulted in a high cost domestic industrial structure that was out of line with world prices.

Hence the new economic policy of 1991 stressed both external sector reforms in the exchange rate, trade and foreign investment policies, and internal reforms in areas such as industrial policy, price and distribution controls, and fiscal restructuring in the financial and public sectors.

In addition, India's membership and commitment to World Trade Organisation (WTO) in 1995 was a clear sign of India's intention to take advantage of globalisation and face the challenge of accelerating its economic growth. One measure of economic growth is given by productivity growth as it forms the basis for improvements in real incomes and welfare.

The concept of productivity growth gained importance for sustaining output growth over the long run as input growth alone is insufficient to generate output growth because of diminishing returns to input use.

This stage, which examines India's productivity growth in the agricultural sector in the context of globalisation, has three main aims. First, it examines these possible links in the agricultural sector in general. Second, it discusses the problems and prospects for agricultural productivity growth of various Indian states. Third, the stage highlights the challenges of globalisation and draws policy implications for the success of Indian agriculture.

OVERVIEW OF INDIA'S AGRICULTURAL ECONOMY

In the early 1950s, half of India's GDP came from the agricultural sector. By 1995, that contribution was halved again to about 25 per cent. As would be expected of virtually all countries in the process of development, India's agricultural sector's share has declined consistently over time as seen in the table below.

Table. Share of Agricultural Output in India's GDP

Year	1950/51	1965	1976	1985	1991	1999	
Percentage	52.2	43.6	37.4	32.8	28.3	24.4	share

The Government's objectives in agricultural policy and the instruments used to realise the objectives have changed from time to time, depending on both internal and external factors. Agricultural policies at the sectoral level can be further divided into supply side and demand side policies. The former include those relating to land reform and land use, development and diffusion of new technologies, public investment in irrigation and rural infrastructure and agricultural price supports.

The demand side policies on the other hand, include state interventions in agricultural markets as well as operation of public distribution systems. Such policies also have macro effects in terms of their impact on government budgets. Macro level policies include policies to strengthen agricultural and non-agricultural sector linkages and industrial policies that affect input supplies to agriculture and the supply of agricultural materials. During the pre-green revolution period, from independence to 1964-1965, the agricultural sector grew at annual average of 2.7 per cent. This period saw a major policy thrust towards land reform and the development of irrigation. With the green revolution period from the mid-1960s to 1991, the agricultural sector grew at 3.2 per cent during 1965-1966 to 1975-1976, and at 3.1 per cent during 1976-1977 to 1991-1992.

Acharya (1998) explains that the policy package for this period was substantial and consisted of:

- Introduction of high-yielding varieties of wheat and rice by strengthening agricultural research and extension services,
- Measures to increase the supply of agricultural inputs such as chemical fertilizers and pesticides,
- Expansion of major and minor irrigation facilities,
- Announcement of minimum support prices for major crops, government procurement of cereals for building buffer stocks and to meet public distribution needs, and
- The provision of agricultural credit on a priority basis. This period

also witnessed a number of market intervention measures by the central and state Governments.

The promotional measures relate to the development and regulation of primary markets in the nature of physical and institutional infrastructure at the first contact point for farmers to sell their surplus products. Acharya (1998) also notes that the rate of growth of productivity per hectare of all crops taken together increased from 2.07 per cent in the decade ending 1985-1986 to 2.51 per cent per annum during the decade ending 1994-1995.

Similar evidence of an increase in yields, a partial measure of productivity gains given by output per unit of land area is seen below for various crops. Although productivity gains were sustained in the 1990s after the liberalisation process began, the yield rates for most of the agricultural products in India are far below comparable rates in a number of other countries. Except for sugarcane, tea, coffee and jute, India's yields are lower than the world average. It should be noted that India is ranked second both in area and output for sugarcane production and is the largest producer of tea and jute in the world.

Although India is doing quite well in wheat production, the average yields in the Netherlands and Ireland are more than three times India's yield rates. In all other major crops, India's productivity performance seems to lag behind others.

Table. Yield for Various Crops (kg/ha)

	1950/51	1960/61	1970/71	1980/81	1990/91	1995/96	1998/99			
Rice	668	1 013	1 123	1 336	1 740	1 855	1 905		Wheat	663
851	1 307	1 630	2 281	2 483	2 596		Coarse	408	528	665
695	900	941	1 035		cereals					
			Pulses	441	539	524	473	578	552	661
	Food grains	522	710	872	1 023	1 380	1 499	1 611		Oil
seeds	481	507	579	532	771	851	948		Cotton	88
125	106	152	225	246	240	Sugarcane	33422	45549	48322	57844
65 395	68 369	69 288								

WHY GLOBALISE

Globalisation in the context of agriculture can be best discussed in the context of three components – improvement of productive efficiency by ensuring the convergence of potential and realised output, increase in agricultural exports and value added activities using agricultural produce, and finally, improved access to domestic and international markets that are either tightly regulated or are overly protected.

These components are linked in various ways. For example, productive efficiency would enhance value added activities in agriculture through agro-processing and exports of agricultural and agro-based products. These activities in turn would increase income and employment in the industrial processing sector. Thus globalising agriculture has the potential to transform subsistence agriculture to comme-rcialised agriculture and to improve the living conditions of the rural community. Ahluwalia (1996) explains that this indirectly requires an improvement in agricultural growth from between 2 and 3 per cent in the past to about 4 per cent per year.

Although initially, with respect to agriculture, there was no major policy reform package in the 1990s, it was however anticipated that the opening up of the agricultural sector to foreign trade, the move to a market determined exchange rate and reduction of protection for industry would, over time, benefit the agricultural sector. Manmohan Singh (1995), the then Finance Minister, in his inaugural address at the 54th Annual Conference of the Indian Society of Agricultural Economics, brought to notice that a policy of heavy protection of the industrial sector operated to the disadvantage of the agricultural sector when industrial prices were raised relative to world prices and thus the profitability of investing in industry was raised relative to agriculture.

This would lead to a shift of resources from agriculture to industry. A policy of heavy industrial protection also led to an appreciation of the exchange rate. Ahluwalia (1996) noted that over-valuation of the exchange rate (before the Indian rupee was devalued by 18 per cent in two phases starting in July 1991) discouraged agricultural exports more than industrial exports because Indian industrial policy had sought to offset the constraints faced by industries via a system of export incentives for market support. Agricultural exports on the other hand were denied any such incentives as they did not use imported inputs.

Ahluwalia (1996) argued that in the past, the agricultural sector was negatively protected because of the above two reasons and the fact that farmers were denied access to the world markets due to trade barriers. Exports of plantation crops and a few commercial crops were free from export restriction but exports of essential commodities, particularly food products, were subject to bans, quotas and other restrictions. Interestingly, Kruger and others (1991) showed that while many developed countries continue to protect agriculture, developing countries do not do so. However, no formal attempt or theoretical framework has yet been used to assess the extent of negative protection in Indian agriculture. The implementation of economic reform in the Indian agricultural sector has been a gradual process. These include an 87 per cent cut in tariff on agricultural products, sustenance of high-yield crop varieties, removal of minimum export price on selected agricultural products, a lift on quantity restrictions on the export of some crops and various land reforms related to tenancy rights and land ceilings.

PRODUCTIVITY GAINS FROM GLOBALISATION AND ECONOMIC REFORMS

In the wake of India's efforts towards globalisation and economic reforms, the expected benefits of total factor productivity (TFP) growth can be represented using the production frontier. The production frontier traces out the maximum output obtainable from the use of inputs.

Opportunities from globalisation and economic reforms can lead to:

- Shift from A to B due to technical efficiency
- Shift from B to C on existing frontier due to input growth
- Upward shift from C to D due to technological progress

Which constitute various sources of TFP growth, can be linked with trade gains. The movement from A to B led by technical efficiency allows increases in output when inputs and technology are used to their fullest potential to obtain the greatest yield. Given that India has been involved in agricultural production for so long, there would be learning-by-doing gains that can help boost production given the expected increase in demand as India opens up.

The increased production would enable a better utilisation of inputs, especially that of advanced capital technology. The reduction in the tariff rate for agricultural products from 113 per cent in 1990-1991 to 26 per cent in 1997-1998 is also expected to motivate local producers into rethinking their production techniques and efficiently utilising the inputs and technology to keep costs of production down in order to remain competitive. The optimum or efficient use of land and water resources would then allow agriculture to respond to the demand for other products such as horticulture and livestock which is expected to increase following a rising trend in the per capita incomes of both rural and urban groups.

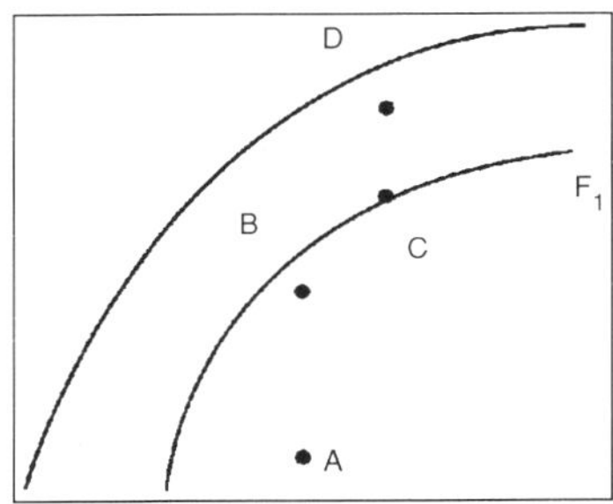

Fig. Total Factor Productivity Gains from Globalisation and Economic Reforms

The move from an overvalued exchange rate to that of a market determined rate would also make agricultural exports cheaper and hence boost exports. The new trading opportunities would necessitate an increased use in the quantity of inputs to boost output and this allows for the movement from B to C along the existing production possibility frontier. Increased exports would bring about economies of scale and as Verdoon's law states, output growth would lead to productivity growth.

The scale of output under increased exports would justify the huge fixed costs underlying technologically advanced equipment and hence increase incentives to adopt high quality inputs. The use of such inputs would result in technological progress and this is represented by the shift from C to D. The reduction in tariff rates in industry from 1990-1991 to 1997-1998 range from 153 per cent to 25 per cent for consumer goods, 77 per cent to 18 per cent for intermediate goods and 97 per cent to 24 per cent for capital goods. This means that farmers now have relatively cheaper access to imported new technology and better capital equipment as well as the option of adopting better farming techniques and this should lead to technological progress.

In particular, the development of agro-processing as an instrument for agricultural and rural modernisation will bring benefits, given its capital-intensive and technology-intensive nature. Lower duty rates on plastics and metals also lower costs of packaging. These forms of cost efficiency should allow competitive pricing of products. In addition, external competition can be expected to motivate local producers into the production of improved quality intermediate inputs for agriculture.

The importance of technology in agricultural development was first demonstrated in the 1970s with impressive growth in yields following the introduction of new wheat and rice varieties. But this technology was limited to areas of assured irrigation as the new seeds also required heavy inputs of fertilizers and pesticides for optimal results.

However, the potential for further extending this technology is not yet exhausted as there is scope for expanding irrigation further and improving the quality of irrigation in many areas. For further technological progress, genetic engineering and the biotechnology revolution provides a prospect of developing new varieties that can flourish with less dependence on water and chemical inputs. Such reduced dependence upon chemical fertilizers and pesticides is also desirable because of environmental considerations, which are an increasing concern.

It must however be acknowledged that the link between trade liberalisation and productivity growth is two-way as they both feed on each other. The productivity gains can be obtained from openness but to benefit from openness via increased demand for exports, agricultural products need to be priced competitively. In other words, productivity growth is necessary to lower the costs of production.

9

Peasant Agriculture

One characteristic of undeveloped peasant agriculture is its self-sufficiency. Farm families in those circumstances consume a substantial part of what they produce. While some of their output may be sold in the market, their total production is generally not much larger than what is needed for the maintenance of the family. Not only is productivity per worker low under these conditions but yields per unit of land are also low. Even where the land was originally fertile, the fertility is likely to have been depleted by decades of continuous cropping. The available manures are not sufficient, and the farmers cannot afford to purchase them elsewhere.

Peasant agriculture is often said to be characterized by inertia. The peasant farmer is likely to be illiterate, suspicious of outsiders, and reluctant to try new methods; food patterns remain unchanged for decades or even centuries. Evidence, however, suggests that the apparent inertia may be simply the result of a lack of alternatives.

If there is nothing better to change to, there is little point in changing. Moreover, the self-sufficient farmer is bound to want to minimize his risks; since a crop failure can mean starvation in many parts of the world, farmers have been reluctant to adopt new methods if doing so would expose them to greater risks of failure.

The increased use worldwide of high-yielding varieties of rice and wheat since the 1960s has shown that farmers are willing and able to adopt new crops and farming methods when their superiority is demonstrated. These high-yielding varieties, however, require increased outlays for fertilizer, as well as expanded facilities for storage and distribution, and many developing countries are unable to afford such expenditures.

THE LABOUR FORCE

As economic growth proceeds, a large proportion of the farm labour force must shift from agriculture into other pursuits. This fundamental shift in the labour force is made possible, of course, by an enormous increase in output per worker as agriculture becomes modernized. This increase in output stems from various factors. Where land is plentiful the output per worker is likely to be higher because it is possible to employ more fertilizer and machinery per worker.

LAND, OUTPUT, AND YIELDS

Only a small fraction of the world's land area—about one-tenth—may be considered arable, if arable land is defined as land planted to crops. Less than one-fourth of the world's land area is in permanent meadows and pastures. The remainder is either in forests or is not being used for agricultural purposes.

General Relationships

There are great differences in the amount of arable land per person in the various regions of the world. The greatest amount of arable land per capita is in Oceania; the least is in China. No direct relationship exists between the amount of arable land per capita and the level of income; Europe has almost as little arable land per capita as Asia and less than Africa; Japan and the Netherlands have very limited amounts of arable land per capita.

The relationship between land, population, and farm production is a complex one. In traditional agriculture, where methods of production have changed little over a long period of time, production is largely determined by the quality and quantity of land available and the number of people working on the land. Until the early years of the 20th century, most of the world's increase in crop production came either from an increase in land under cultivation or from an increase in the amount of labour used per unit of land.

This generally involved a shift to crops that would yield more per unit of land and required more labour for their cultivation. Wheat, rye, and millet require less labour per unit of land and per unit of food output than do rice, potatoes, or corn (maize), but generally the latter yield more food per unit of land.

Thus, as population density increased, the latter groups of crops tended to be substituted for the former. This did not hold true in Europe, where wheat, rye, and millet expanded at the expense of pasture land; but these crops yielded more food per acre than did the livestock that they displaced.

As agriculture becomes modernized, its dependence upon land as well as upon human labour decreases. Animal power and machinery are substituted for human labour; mechanical power then replaces animal power. The substitution of mechanical power for animal power also reduces the need for land. The increased use of fertilizer as modernization occurs also acts as a substitute for both land and labour; the same is true of herbicides and insecticides. By making it possible to produce more per unit of land and per hour of work, less land and labour are required for a given amount of output.

Recent Trends

Crop yields have increased dramatically since 1950, with a faster rate of growth in the developing than in the developed countries. Most of this increased output has been due to gains in yields rather than to the expansion of cultivated land. In Europe as well as in North and Central America, the total area under crops has declined; in South America it has increased by more than one-half and in Asia by more than one-third. The large increase in Oceania was due to immigration. The large decrease in Africa was due to a succession of droughts from the 1970s on. Grain yields in the developed regions of the world have increased consistently over the past several decades. In the rest of the world the pre-World War II yields were not achieved again until the mid-1950s. The increases in grain production were more than twice as high in the developing as in the developed countries.

Food production and total agricultural production exhibit nearly identical trends, and changes in food production can be taken therefore as indicative of changes in total agricultural production. Food supplies per capita in developing countries have increased at nearly the same rate as in developed countries, indicating a narrowing gap between food supplies and population growth in the developing countries.

ELEMENTS OF COMMERCIAL AGRICULTURE

Although elements of the above ideal type of a competitive food system could be found in parts of the United States at various times, it never did exist in many areas of the country. For an individual farmer, the question was not how many firms were involved in the different stages of the food chain across the country, but rather how much of this commodity chain was accessible to his (sometimes her) farm. AS farmers moved west, one of their major problems was how to transport their products to markets in the eastern cities.

The government, wishing to promote increased industrialization, also perceived the problem and subsidized the construction of transportation systems, especially railroads. This often made the farmer dependent on a monopoly which could exploit him/her because of the unequal balance of economic power. If a farmer had access to only one railroad, the power relationship certainly favored the railroad. That farmer faced a monopoly regardless of how many other railroads existed in the country. Thus, many farmers faced the issue of monopoly control of capital from the time they became commercial farmers and began to be dependent on a single transportation system to move their products to the market.

Railroads in some parts of the country needed the business of farmers, but they had access to hundreds, if not thousands, of farmers. They were not dependent on any single farmer. Their only concern was that farmers might be successful in organizing a united stand against the railroads. Frank Norris' agrarian populist novel, The Octopus: A History of California (1901), was about a handful of farmers who attempted such a united stand, rising up in direct rebellion, only to be dispatched by the railroads. The whole history of the farmers' movement is largely about the unequal power balance between farmers and the railroads, and, more generally, between the farmers and all the "middlemen" they depended on for transportation, markets, and a host of inputs such as credit and farm equipment.

In southwestern Minnesota, for example, the selling of water fowl to cities to the East and later the selling of agriculture products, especially grain, all depended on the railroad for transportation. In the case of grain, the farmers also depended on the elevators (large silos) to store and to transfer it from farmers' wagons to the train. Many of the transnational corporations (TNCs) of today, like Cargill, exercised economic power in many of the local markets in which they began operating. Most farmers' movements were not successful in establishing alternative economic systems or firms which benefited the farmers, but with the help of legislation to encourage farmer cooperatives, there were some successes.

Advantages of Cross-subsidization to Concentrated Capital

In the food system, horizontal integration usually refers to expansion in the same stage of the same commodity sector. However, if one considers the sector to be meat, then horizontal integration would include the total meat sector. For example, ConAgra ranks in the top four firms in the processing of beef, pork, broilers, sheep, turkeys, and seafood (which is not on the list). Spokespersons for the industry frequently highlight the competition for the public's dollar between different meats, such as the competition between beef and poultry.

They frequently use the competition between meats to argue that the producers must make certain changes in their practices. This competition between commodities is also frequently used to justify the check-off system in which a per-animal fee paid by farmers is used primarily to support product promotion and research. The cost of the check-off-system is borne by the producers, but there is growing concern as to who is benefiting.

Currently the Livestock Marketing Association is leading a petition drive to force a beef producer recall referendum on the check-off programmes. Their argument is that "after $1 billion spent in promotion and research over the first 10 years of the programme, beef demand is still declining".

They suggest the check-off funds are being spent directly or indirectly on projects that benefit processors and retailers rather than beef producers, and that the results are increased concentration and integration of the industry. One can ask how much competition exists between the different meat products when key decision-makers are involved in more than one part of the meat sector.

The movement of firms into the processing of several commodities may at one level be an extension of horizontal integration, but it also represents a major qualitative change in the economic power relationships. When a firm has a dominant position in several commodity systems, it can cross-subsidize.

Firms operating in more than one commodity system gain economic power because they can survive a major loss in one commodity system over a long period of time if they are making significant profits in other systems. If a loss continues very long in a single-product firm's only commodity, it faces serious financial difficulty.

Lane Poultry was the largest broiler producer and processor in the United States following its purchase of Valmac Industries in 1980, but it was still a single-product producer. Lane lost Valmac Industries and then was itself purchased by Tyson Foods because of its economic losses in nine of eleven consecutive quarters in the late 1970s and early 1980s. Being the largest firm in a commodity sector, but a single product producer, does not assure enough economic power to survive. Larger firms with profits in other sectors or systems have more economic power and may overtake them. Information we obtained from executives of a couple TNCs involved in broiler production at the time indicated that the goal of their firms was to obtain a larger share of a growing market.

Planned overproduction and selling below cost of production also occurred in the farm-raised catfish sector early in the 1980s. Two major catfish cooperatives, Southern Pride and Delta Pride, experienced the problem of competing against ConAgra, Cargill, and Chiquita, the parent company of Morrill (now owned by Smithfield). The three TNCs were able to cross-subsidize. The cooperatives survived despite the fact that the annual report of one of the TNCs showed a loss in the catfish division for two years because of "overproduction" in the sector.

Conversations with some of the TNC's personnel indicated the firm was prepared to extend this loss for another year or two. We concluded that because the five firms absolutely dominated the production and processing of farm-raised catfish at the time, the low prices were the result of overproduction and an effort on the part of the TNCs to gain market share at the expense of the members of the cooperatives. In fact, Cargill, which has now exited the sector, entered the catfish sector during the time of negative profits with plans to increase catfish production.

Like the broiler and catfish sectors in the past, the hog sector is currently involved in a large increase in production even when prices are low and are predicted to stay low. The issue is market share, not efficiency. Large firms that can cross-subsidize can operate in this arena, but smaller, nondiversified firms cannot survive. Economic power, not efficiency, predicts survival.

Horizontal Integration

Most food firms started as relatively small, local firms, but as they became profitable they expanded their operations into other geographic areas. The expansions occurred through building new facilities, acquisitions, and mergers. The expansion of a firm within the same stage of the food system as their original operation is called horizontal integration. For example, the increase in size and decrease in the number of farms in the United States during most of this century is an example of horizontal integration. Horizontal integration also occurs at each of the other stages of distribution and processing. Although there are great variations among the different commodity sectors regarding the ways concentration of ownership and control have occurred in processing and distribution stages, the same general pattern of fewer and larger firms in each stage has been underway during the last half of the twentieth century in the United States and has become most obvious during the last decade at the global level.

In some commodity sectors, one can point to significant concentration of the processing firms in even the first half of the century. For example, pork and beef slaughtering and processing were dominated by Wilson, Armour, and Swift as we entered the twentieth century. Opposition to their practices in the Chicago stockyards inspired Upton Sinclair's The Jungle (1906) and led to the passage of the first Food and Drug Act that same year. And their collusion to set monopoly prices was largely responsible for the creation of the Packers and Stockyards Agency of the U.S. Department of Agriculture (USDA) in 1921 to monitor predatory practices. The Swift and Armour brand names exist today, but the firms were bought by ConAgra, which also bought Miller and Monfort. Some would argue that the fact that these firms do not exist today suggests that even firms with significant economic power can themselves be eliminated. But the important point is that they were absorbed within the larger agglomeration of capital. Still it is true that with the continuing trend towards concentration and centralization of capital no firm is safe from takeover or elimination in other ways.

Forty percent or more of the processing of all agricultural commodities in the Midwest are controlled by the four largest firms. Although debate continues in the United States and in other countries on what constitutes an oligopolistic or near oligopolistic market, much of the economic literature suggests that when four firms control 40 percent of the market, they are able to exert influence on the market unlike that in a competitive system. In the meat sectors 87 percent of the beef cattle are slaughtered by the four largest firms (81 percent by the largest three) and 73 percent of the sheep are processed by the four largest firms. The control of hog slaughtering by the four largest firms increased from 37 percent in 1987 to 60 percent today. Over one half (55 percent) of the broilers (chickens produced for meat) today are produced and processed by the four largest firms, with Tyson now producing and processing almost one third of the broilers in the United States. In the crop sectors, the four largest firms process from 57 percent to 76 percent of the corn, wheat, and soybeans in the United States.

Like the narrow opening of an hour glass which controls the flow of sand from the top to the bottom, the processing firms are positioned between the thousands of producers and millions of consumers in the United States and the world. These firms have a disproportionate amount of influence on the quality, quantity, type, location of production, and price of the product at the production stage and throughout the entire food system. The only stage in which a set of firms begins to equal the economic power of the food processors is the retail stage, which is also becoming more horizontally integrated. The interface between the processing and retail stages is currently where the giants of the food system interact. Certainly, it is not an area characterized by easy entry and exit. How many firms in the world have sufficient capital to face the economic power of these two sets of firms?

Vertical Integration

The second major strategy of monopoly capital is vertical integration. Vertical integration occurs when a firm increases ownership and control of a number of stages in a commodity system. Just like diversifying into different commodity sectors (horizontal integration), this structure gives the firm more economic power.

Another example of the extent of vertical integration in the food system comes from ConAgra's annual report. ConAgra indicates that it is the largest distributor of agricultural chemicals in North America, one of the largest fertilizer producers, and in 1990 it entered the seed business. (Since then it has formed a joint venture with DuPont and formal relationships with some of the seed companies involved in biotechnology). ConAgra owns 100 grain storage elevators, 2,000 railroad cars, and 1,100 barges. ConAgra is the largest turkey producer and second largest broiler producer. It produces its own poultry feed, as well as other livestock feed. It also owns and operates hatcheries. ConAgra hires growers to raise its birds and then it processes the birds in its own facilities. This broiler meat can then be purchased as fryers under the name of Country Pride or in further processed foods such as TV dinners and pot pies under the labels of Banquet and Beatrice Food. From the basic raw materials for agricultural production to the retail store, a significant proportion of the food system is owned and controlled by ConAgra. ConAgra is the second largest food firm in the United States (behind Philip Morris) and the fourth largest in the world, with operations in thirty-two countries.

In the subsistence food system, the family controlled its food from seed to plate. In the emerging vertically integrated food system, a few food companies are gaining control of the country's food system by controlling it from seed to shelf. This system is being extended around the world by many of the firms that are headquartered in the United States.

Starting in the 1950s and 1960s, when feed companies and others started hatching their own baby chicks, hiring growers to provide labour, buildings, and land, and constructing their own processing facilities for broilers, the farm press and farm community began to focus on contract production. Contract production is very different from forward contracting of a commodity product. Forward contracting is a sales agreement between a farmer and a buyer that involves an agreed upon price and other terms of the sale to be carried out at some future date. Contract production is an industrial model in which the integrating firm outsources a needed ingredient-the agricultural raw product. In contract production, the growers are required to provide the land and the buildings, and equip the buildings to the integrating firm's specifications while providing all of the labour for the production stage of the system.

The growers are thus hired workers paid on a piece rate basis. They never own the birds or the feed, and have no knowledge of the genetics or the feed ration. The integrating firms provided the birds, feed, and medication. All of the major decisions are made by the integrating firm. The growers mortgage their land to raise the capital to build the buildings, which cost over $100,000 each. Typically their repayment schedule extends over a ten-to fifteen-year period, while the contract with the integrating firm goes from one batch of chickens to the next-a period of about six weeks. By the time the buildings are almost paid off, the equipment needs to be replaced and the buildings need to be modernized. As a consequence, few growers ever get out of debt. It is estimated that although about one half of the capital in the broiler sector comes from the growers, all of the major decisions are made by the integrating firms. The growers are well aware that they can be cut off at any time.

In the early stages of vertical integration in the broiler industry, most growers had access to several integrating firms, but over time the numbers were reduced. For example, in 1969 in Union Parish, Louisiana, there were four integrating firms. Two were locally owned feed operations, and two were operations based out of state. By 1982, the two local firms were no longer integrating firms; they were now growers. The two outside firms were owned by ConAgra and Imperial Foods, one of the largest food companies of England. Within the following year, ConAgra bought the Country Pride broiler facilities from Imperial Foods. The growers report that they have had no price increase since 1982.

Two processes occur that alter the growers' opportunities. First, as the integrating firms in a given geographic area become fewer, the power relationships between growers and the firms become more unequal. Because of transportation and other costs, most integrating firms will send trucks out only about 25 to 30 miles from the processing site to deliver feed and to pick up poultry for processing. Today there are about 40 firms producing about 97 percent of the broilers in this country. In total, they operate about 250 processing facilities. Thus, there are very few growers who live close enough to more than one processing facility to even have an option to choose between integrating firms.

The second process that limits options for growers is that as the number of firms operating in the same geographic area declines to two or three, an informal agreement evolves between integrating firms that they will not raid their competitors' growers. If a grower gets cut off from one integrating firm, they cannot enter into a contract with another. My thirty years of observing the poultry sector suggest that in early capitalism when the growers have access to several integrating firms, the growers experience financial success, but when the system moves to monopoly capital the growers find themselves in financial crisis. The courts have also found that the growers are at the mercy of the integrating firms in other ways. Errors in the weighing of both feed and poultry have become so well documented that legislation has recently been introduced to address this issue. The USDA has also made a commitment to study this problem.

ROLE OF FORESTRY SECTOR IN INDIAN AGRICULTURAL ECONOMY

India, in spite of having 2.5% of the world's geographic area and 1.8% of the world's forests, sustains 16% of the planet's human population and 18% of its livestock population. Forestry contribution is 1.7% of nation's GDP. This does not take into account unrecorded withdrawals (NWFP, fuel, wood, fodder etc.). Moreover, the environmental benefits of forests also remain to be quantified and calculated.

The forests contribute 1.7% to the GDP of the country (NFAP 1999 a & b). Due to problems associated with the valuation of forests and services, unrecorded removals, illegal harvesting, etc. the exactness of the contribution has not been established. A large part of the forest production consisting of fuel, fodder, medicine and food are removed without payment and without any record by the rural and tribal people. According to Ahmed (1997), the total annual value of India's harvest of all forest produce is estimated to be Rs. 300,000 millions (compared to the investment of Rs. 8000 in the sector). The low estimate of contribution to the GDP resulted in low priority for forestry investments in five year plans. Efforts are needed for monitoring the services provided by the forests so as to appreciate their contribution to human well being. Over 50% of the revenue earned by the forest departments comes from NWFPs. Their growth is generally 40% higher than timber (MOEF 2000). Nearly 350 million people living in and around forests in India depend on NWFPs for their sustenance and supplemental income which is worth Rs.400 billion annually (Tewari 1994). Studies in Orissa, Madhya Pradesh, Himachal Pradesh and Bihar have indicated that over 80% of forest dwellers depend entirely on NWFPs.

Similarly 17% landless depend on daily wages related to the collection of NWFPs. 39% people are, however, involved in NWFPs collection as a subsidiary occupation. It has been estimated that many village communities derive as much as 17-35% of their annual household income from the sale of NWFPs. NWFPs provide 50% of the income to about 30% rural people. The average income realized through the sale of NWFPs by households in the state of Madhya Pradesh constituted 34 to 55 percent of their total income. As per estimates made in West Bengal, an average return of Rs 2270 ha yr-1 is obtained from NWFPs, which is 25% more than the polewood harvest, which fetches Rs 16,000 per ha after 10 years. There is, thus sufficient evidence to believe that the collection of NWFPs is a crucial part of the population's life support system, especially of the tribals.

Growing Pressure on Forests

The present ecological conflicts have created many economic compulsions and sociological stresses due to changing consumption patterns, scarce availability of land and other natural resources. Out of the total requirement for wood, 70% is for fuel wood and 30% for timber. Thus, forests have at least 5 times more pressure than what they can withstand. This is in addition to the 30% contribution to the fodder requirement of the country in the form of 178 million tonnes of green fodder and 145 million tonnes of dry fodder.

Employment Generation

Of the total wage employment in the forestry sector, NWFPs account for more than 70% of the opportunities for self-employment for the forest dwellers as farm mechanization has not developed well in India. According to an ILO estimate, one hectare of forest plantation creates nearly 630 mandays, from the raising of nurseries to the harvesting stage. 70% of the budget allocated to plantations or afforestation is spent on providing direct wages to the workers and only 30% goes towards purchase of seeds, planting materials, equipment etc. It would not be out of place to mention that 50% of the workforces on forest plantations are women and tribal. Rural women use 70-80% of the mandays in collection of NWFPs, fuel and fodder.

Activities related to NWFPs provide employment during slack periods and a buffer against risk and household emergencies. In the remotest areas, sometimes the forest is the only source of employment and income. Research is needed to evolve forest based entrepreneurial endeavours to produce multiplier effects through the forward and backward linkages.

Markets

The dependence of the producer on intermediaries and his limited access to markets has a direct effect on prices. The price of a product whether sold to consumers directly or through intermediaries, has no bearing on the expenditure incurred on the labour, inputs and transportation. Under direct sales, the localized activity for localized markets creates a supply position in excess of local demand. Traders control the market and dictate the prices during the season and in the off-season. The sale of produce during the flush season and in the off-season is different. In the case of sale through intermediaries, the producers have absolutely no control over the prices. Studies show that the poor primary producer's income always remains low. The need for market related studies has always been felt, and includes research on the market information system and the scope of value addition like bioprospecting, Intellectual Property Right (IPR) etc.

MOVEMENTS AAND SUSTAINABLE DEVELOPMENT

Most of the recent discussion of 'environmental management' in developing countries has paid scant attention to the role of local people. Environmental management is defined almost exclusively terms of a rigidly 'top-down' approach. It is assumed that by incorporating better systems of environmental accounting in the national accounts, and by ensuring that the negative costs of development are allowed for within project budgets, progress will be made towards better management of the environment.

Desirable as it is that economics takes environmental problems seriously, such measures are unlikely to affect the balance of forces which cause poor rural people to place excessive strain on their environments. The 'official' discussion of environmental management in the South makes barely a reference to the increasingly important experiences, and literature, on environmental movements there. The remaining section of this chapter focuses attention upon the need to re-examine our knowledge and understanding of the environment, in the light of the experiences of other cultures far removed, socially and geographically, from the 'core' of the modern food system.

SUSTAINABILITY AND DEVELOPMENT

It is clear that rural development projects have often neglected local, ecological and cultural factors in the South. It is also clear that structural factors—which influence people in not acting 'sustainably'—have frequently been ignored. We need, in fact, to take account of both the cultural and structural dimensions in any definition of sustainability. Then we can 'begin to ask whether there are patterns of myth and social organisation, which are more likely to ensure and maintain man's *[sic]* survival base'. It is clear that environmental problems are not new. Historical, and archaeological, evidence suggests that early societies often behaved in ways which were damaging to their environment.

These societies also had to 'solve the basic problem of producing food surpluses and collecting raw materials from rural areas to sustain large urban populations', a problem that faces countries in the South today. However, such societies, unlike the South today, were not faced by twin assaults from rapid economic and demographic growth, on such a scale. They did not exist within a global context which placed their own food systems and values in jeopardy.

If we are to understand fully the relationship between sustainability and development, we must begin by recognizing that they are paradigms with historical as well as geographical antecedents which need to be located within specific social and economic contexts. The point at issue is that societies differ in the way they prescribe courses of action towards the environment. Too often we assume systems possess universalistic characteristics, including the food system.

Such universalistic approaches to social and political systems often leave no room for human agency within the framework of explanation. The question we might address in exploring environmental movements was posed by Giddens in the following way: how does the 'practical consciousness' that people acquire from their society, and from their environment, relate to the 'structural properties of the system'? Environmental movements include movements of resistance to the encompassing food system. Under what circumstances are the structural properties of such a system called into question?

THE EPISTEMOLOGICAL DIMENSION

To answer this question we might begin, not with an analysis of political conflict over the environment, but with the way in which the 'environment' is understood. Instead of working 'backwards' from social movements to consciousness, we can work 'forwards', from the formation of environmental consciousness to the organization of alternative forms of social action to redress environmental problems.

It is clear that the way in which knowledge is gained about the environment and the way in which this knowledge is used have considerable bearing on the question of human agency, and with it environmental management practices. Howes has argued that in agricultural systems engaged in monocrop cultivation the world view is one in which 'society is conceived as an entity apart from Nature'. It is clear, too, that many non-Western cultures possessed 'scientific theory' before Western science, as we know it, had fully developed. Looking at examples from South Asia, Goonatilake claims that, 'their large-order cognitive maps anticipated the non-manipulable, and distant reality of modern science'. Furthermore, these kinds of knowledge system often codify knowledge in ways that ensure its cultural survival, long after the reasons for certain observances have passed. In some cases 'after an initial map has been made and legitimized...it tends to continue to be believed for considerable periods of time'. A useful distinction, although it should be emphasized that it exists *within* societies as well as between them, is between what Feyerabend calls *historical traditions* and *abstract traditions.* The two traditions can coexist in a society at any point in time, but some cultures give more emphasis to one tradition than to the other.

Historical traditions involve a different discourse from abstract traditions. It is assumed that 'the objects already have a language of their own' and it is the task of people within a society to learn this language. Their task is to 'learn the language of the objects as they are, and not as they appear, after they have been subjected to standardizing procedures' and experimentation. According to this tradition there is no such thing as pursuing 'objective truth', since truth involves both an object and a subject. Feyerabend compares this with what he terms an abstract tradition.

Abstract traditions are different in that they involve framing statements as a means of building up a model of what happens in nature. According to Feyerabend, 'the statements are subjected to certain rules (of logic, testing and argument) and events affect the statements only in accordance with the rules' It is possible, then, to make scientific statements without personal contact with the objects being described. Both these epistemological traditions exist, at different levels, in most societies, but within the Western scientific tradition much more emphasis is given to abstract traditions.

In this respect the development of Western science is best understood in the context of Judaeo-Christian thinking, which has often appeared inimical to 'scientific' method. For example, Goonatilake argues that religious and philosophical objections to scientific method in Europe would not have occurred in South Asia, where most religious beliefs emphasize the role of both abstract and historical traditions in arriving at valid statements about the world around us. The first point that needs to be emphasized in establishing a framework for understanding environmental conflict in the South is that a society's epistemological framework will influence its understanding of the environment. What we perceive as 'environmental' problems may be interpreted in other cultures as essentially due to the breakdown of social and cultural practices and beliefs.

In addressing these questions we need to be aware that our understanding of Societies very unlike those of the industrialized countries frequently distorts our understanding of their environments. By the same token, the rediscovery that knowledge-systems other than our own have importance for understanding our relationship to the environment has become a central, but contentious, tenet of recent Green thinking. As Luke observes, some Green thinkers have misunderstood the importance of non-Western science. 'Deep ecologists' have sometimes equated small-scale, non-Western societies with 'primal societies', implying that 'primitives' provided an echo of our own pre-Enlightenment past.

This rather naïve view of the superiority of non-Western epistemologies fails to recognize that myths, magic and ritual are 'the functional equivalents (and perhaps the conceptual antecedents) of Enlightenment science and technology'.

This kind of simplification and misunderstanding also errs in lifting ecophilosophies out of their cultural context—Taoism or Buddhism—'with little consideration of any cultural grounding', when it is the specific cultural context of environmental epistemologies that gives them their force and vitality. As an approach, the close association of indigenous knowledge with 'answers' to current environmental problems is frequently ethnocentric, a historical and reductionist, substituting descriptions of the world for more rigorous and less ethnocentric analysis.

Other examples exist of societies closer to our own in which environmental problems are constructed from very different political and social milieux. In Eastern Europe today, for example, the commitment to radical political ecology, and the growth of Green movements, is also linked to processes that are not strictly 'environmental' in the narrow sense employed in the West Conflicts over ethnic identity, and the struggles for civil and human rights in contemporary Eastern Europe, have incorporated—and adapted to—environmental demands, and often express these demands in language drawn from the discourse.

THE ECONOMIC DIMENSION

The second, important dimension of relations between the environment and society is the economic dimension. The environment is subjected to structural changes over time, linked to economic growth and the development process. If we confine ourselves to the relationship between agriculture and the environment we can identify several features of the development process which have major implications for environmental systems, among them: the production of food as a commodity rather than for household consumption; the adoption of 'high-technology' energy systems on the farm, and in the food processing industries, rather than 'low-technology' energy inputs; and the role of technology in reducing the pressure exercised by population increase on limited natural resources. These factors will be looked at in turn.

FOOD AS AN EXCHANGE VALUE

In some societies agricultural practices underwrite sustainability, in the sense that consideration of the long-term implications of current agricultural practices is enshrined in farm management practices. For example, Wilken shows how some small farmers in Central America, without recourse to fossil fuels, effectively 'bank' their own labour, and that of their household, investing labour in resource management practices that make sense within a longer time-frame. They are seeking to reduce future operational costs on the farm by making major current investments in labour-intensive management now.

As we have argued above, in some societies such practices become ritually encoded. Rappaport shows how, under stable conditions, the ritual enactment of myths may not only help to maintain the social cohesion of the group (a point often made by anthropologists) but also help to sustain the natural environment. Of course, the value placed on a resource is linked to other social referents, such as the concept of ownership, as well as economic factors like the cost of extraction.

The development of wider markets for food products, and the adoption of production techniques that depend on sophisticated off-farm processing and industrial inputs, does not necessarily mark the point at which food becomes a commodity—in most societies food has been a commodity for a long time. However, it does mark the point at which the conversion of food from a use value to an exchange value has the greatest impact on the way the environment is managed, and ultimately socially constructed. Food is taken from the province of reproduction in small-scale societies and transferred to that of production and consumption in more complex societies. Eventually, in what we can term 'post-industrial societies', the locus of dissent shifts again, this time from production relations to those of consumption. Some of the most important implications for the evolution of the food system are represented by resistance to the introduction of new, processed 'convenience' foods.

FOOD AND ENERGY TRANSFERS

The food system also affords an excellent example of the way that energy use is linked to the capacity of ecosystems to retain, as well as transfer, value. The process of development, as we have seen, is usually considered in relation to the movement of capital and labour, rather than the exploitation of natural resources. The structural and spatial aspects of transfers of value from one part of the (developing) world to another part of the developed world have been emphasized by O'Connor and Smith. Without dissenting from the significance of these forms of surplus extraction, it is important to emphasize that transfers of energy from within one system to another can also be seen as transfers of value.

Bunker uses both ecological principles, and the laws of thermodynamics, to assert that 'our calculus of value must include not just the labour and capital incorporated into commodities, but all forms of energy and matter'. In his view, 'the embodiment of energy in economic and social organization encompasses far more of the essential differences, and relations, between core and periphery, than measures linked to commodity production and exchange can'.

POPULATION INCREASE AND CARRYING CAPACITY

Economics has also experienced considerable difficulty recently in incorporating the full implications of changes in environmental systems within its analytical framework. The Malthusian tradition emphasized that population pressure placed limits on the capacity of natural resources to feed the population. In fact, like many economists, Malthus did not have a great deal to say about the environment. Barbier reminds us that the pessimism of the Malthusian tradition was only matched by the optimism of the Ricardian view of scarcity. This placed emphasis on 'relative scarcity' rather than 'absolute scarcity'; as resources became used up substitutes were found for them, extending the time-frame in which the environment could facilitate economic growth.

The problem of carrying capacity was not solved but it was 'put back', and technology was the major factor in effecting this change. Later neo-classical thinking, represented by Alfred Marshall, took an even more optimistic position than Ricardo; no tendency was predicted for profits, wages and rent to decline with increased scarcity of natural resources. The implication was that we were entering an 'age of substitution', the environment could be exploited without degradation being inevitable, indeed the scarcity of non-renewable resources might attract a higher price, and therefore lead to their better conservation. This view of 'substitution' still plays a large part in the way we conceptualize the environment today, and has been given added impetus by developments in biotechnology.

The importance of the economic dimension is that it necessarily involves considering the conversion of natural capital into social capital, economic processes of growth and development do not simply modify, they actually transform the environment. In the course of these transformations the development process 'hands back' to nature some serious problems, including that of sustaining economic levels of production and consumption.

AREA STATISTICS IN AGRICULTURE

India is primarily an agriculture-based country and its economy largely depends upon agriculture. Presently, contribution of agriculture about one third of the national GDP and provides employment to over seventy percent of Indian population in agriculture and allied activities. Therefore, our country's development largely depends upon the development of agriculture. The agricultural production information is very important for planning and allocation of resources to different sectors of agriculture. Agricultural statistics in India have a long tradition. Artha Shastra of Kautilya makes a mention of their collection as a part of the administrative system.

During the Moghul period also some basic agricultural statistics were collected to meet the needs of revenue administration. Ain-e-Akbari is most important document which throws great light on the manner in which statistics were collected during the moghul period. After the Moghul period British rule started in the country Ryotwari System was introduced during 18th Century by the East India Company to collect land revenue.

The historical famine of 1860 emphasised the need for more statistical information. In 1866, the British Government Initiated collection of agricultural statistics mainly as a byproduct of revenue administration and these reflected the then primary interest of the Government in the collection of land revenue. Subsequently, the emphasis shifted to crop forecasts designed primarily to serve the British trade interests. On a representation made by a leading firm of Liverpool, trading in wheat, the preparation of wheat forecast was taken up in 1884 and the land utilisation statistics are available in the country since 1884. By 1900, oilseeds, rice cotton, jute indigo and sugarcane had also been added to the list of forecast crops.

After the First World War significant improvements were made in the agricultural statistics of the country. The Royal Commission of Agriculture was appointed in 1926 by Government of India, to examine the conditions of agricultural and rural economy. The report of the commission was published in 1928. Considerable improvements in the statistics collected were brought about through acceptance by the Government, the recommendation made by the Royal Commission on Agriculture (RCA), with regard to the quality and coverage of the statistics as well as for re-organising the country's statistical set up. The Commission recommended the constitution of the Imperial Council of Agricultural Research which was renamed after independence as the Indian Council of Agricultural Research (ICAR).

During the Second World War, when the attention of the Government was focused on the critical food situation, the need for timely and reliable statistics of food production was keenly felt for implementation of food policy and administration of controls. The initiation of the crop-cutting experiments based on random sample surveys for estimation of yield rates of principal crops for replacing the traditional eye-appraisal method was the direct result. In 1949, the Technical Committee on Coordination of Agricultural Statistics (TCCAS) was set up by the Ministry of Agriculture which highlighted the gaps in agricultural statistics and the improvements necessary to remove the defects.

In an attempt to fill the gap, during that period Prof. Mahalanobis introduced a new statistical system to estimate crop statistics which is known as Grid sampling. In sample surveys the final estimate is prepared from information collected for sample units of definite size (area) located at random. In large-scale surveys, cost and precision of the result depends on size of sampling units (area) and the number of sampling units. Therefore, it is important to strike a balance between these two quantities in planning of surveys. In this context, the approach of grid sampling has been proposed by Prof. P. C. Mahalanobis for areal sampling. The whole area is considered as a statistical field consisting of a large number of basic cells each having a definite value of the variate under study. These values (with suitable grouping) form an abstract frequency distribution corresponding to which there exists a set of associated space distributions generated by allocating the variate values to different cells in different ways. This raises novel problems which are space generalisations of the classical theory of sampling distribution and estimation.

On the applied side it also enables classification of the technique into two types:

1. 'Individual' or
2. 'Grid' sampling depending on whether each sample unit consists of only one or more than one basic cell.

For most space distribution, precision of the result is nearly equal for both types of sampling; these are called fields of random type. For certain fields (including those usually observed in nature) precision depends on sampling type, *i.e.,* these are fields of non-random type. This technique was applied in estimation of acreage under jute covering 60,000 sq. miles in Bengal in 1941-42, and it was observed that the margin of error of the sample estimate was about 2 per cent, while cost was only a fifteenth of that of a complete census made in the same year by an official agency.

With the ushering in of the planning era in 1951-52 greater attention was paid to the improvements in the collection of statistical data on a number of items and various schemes for improvement of agricultural statistics were implemented by the Central and State Governments as part of the successive five year plans. During the First and Second Plan periods, the Directorate of Economics and Statistics (DES), Ministry of Agriculture and Irrigation sponsored schemes for adoption of basic annual and quinquennial forms recommend by the TCCAS, extension of reporting area, estimation of production of principal foods and minor crops of commercial importance, rationalised supervision over the work of the area enumeration, preparation of index numbers relating to agricultural economy, etc.

A number of other organisations like the Institute of Agricultural Research Statistics (IARS) presently named as Indian Agricultural Statistics Research Institute (IASRI), Central Statistical Organisation (CSO), Directorate of National Sample Survey, now re-organised as the National Sample Survey Organisation (NSSO) and the Indian Statistical Institute (ISI) also participated in the efforts directed towards the improvement of agricultural statistics in different ways. A standing Committee on Improvement of Agricultural Statistics (CIAS) was set up in the Ministry of Agriculture in 1961 to guide and review the implementation of schemes for improvement of agricultural statistics.

Efforts for improving the quality and content of the statistics continued during the Third and Fourth Plan periods. In December 1969, a Data Improvement Committee was set up under the chairmanship of Dr. B.S. Minhas to look into the problems of improving the data base of the economy. In regard to agricultural statistics, this Committee pointed out the gaps in data and made several important recommendations for improvement to meet short term policy needs. In retrospect, the decade of fifties witnessed a period of initiation of new schemes for improvement of agricultural statistics, while during the sixties, efforts were made to consolidate the improvements.

The National Commission on Agriculture was appointed in 1970 under the chairmanship of the then Minister of Agriculture and Irrigation, Govt. of India which made several important recommendations for strengthening and improving the system of data collection. Agriculture being a state subject and statistics falling in the concurrent list the Agricultural Statistics System is a decentralised one with the state governments (The State Agricultural Statistics Authorities, or SASAs, henceforth) playing a predominant role in collection and compilation of agricultural statistics and more particularly the crop statistics.

The Directorate of Economics and Statistics (DES), Ministry of Agriculture at the Centre is the pivotal agency for coordination and compilation of agricultural statistics at all India level. Other principal agencies which collect data and conduct methodological studies on agricultural statistics are the National Sample Survey Organisation (NSSO), the Indian Agricultural Statistics Research Institute (IASRI), the State Directorate of Economics and Statistics (State DESs),etc. The present system of agricultural statistics generates valuable statistics on a vast number of parameters. Some of the very important statistics are land-use statistics and area under principal crops through the Timely Reporting Scheme (TRS) and also on complete enumeration basis, yield estimates through the General Crop Estimation Surveys (GCES), cost of production estimates, agricultural wages, irrigation statistics, etc.

It also generates data on livestock products through the scheme of integrated Sample Survey (ISS), collects wholesale and retail prices, conducts market intelligence and observes rainfall and weather conditions. The basic information on various aspects is also collected through the Agricultural Census and Livestock Census on quinquennial basis.

SYSTEM OF DATA COLLECTION FOR AREA ESTIMATION

The country can be divided into four broad categories with respect to collection of area statistics namely:

- Temporarily settled states also known as 'Land Record States',
- Permanently settled states also known as 'Non- Land Record States',
- Other regions and
- Non-reporting areas.

Temporarily Settled States

The system of temporarily settlements was introduced in our country in 1892, with a view to fix land revenue for a period, which was subject to change at the time of the next settlement. Ordinarily, the interval between two settlements was 25 to 30 years. In order to determine the land revenue and to develop estimates of production detail statistics are to be collected about land revenue, land value, etc. In temporarily settled areas the information on crop area statistics are collected by the village accountant or Patwari and are recorded in a register which is popularly known in northern India as Khasra.

The village accountant has been called by different names in different parts of the country such as Karnam in South, Telathi in Maharashtra, Karamchari in Bihar, lekhpal in Uttarpradesh, etc. This category covers around 86 per cent of total reported area of the country. The crop area statistics collected by village accountant in the Temporarily settled States are on the basis of complete enumeration called girdawari. The village accountant is to visit each and every field of the village in each crop season and record the information such as area under different crops/ land use categories and its status in standard forms called Khasra register.

The work of village accountant is supervised by immediate superior officer known by the name of Quanungo in northern India. Most of the geographical areas of temporarily settled states are cadastrally surveyed and detailed maps are available in tehsil and district Headquarters. The statistics obtained by different village accountants are aggregated to get the crop area statistics at higher administrative units such as blocks, tehsils, district, states, etc.

This system of data collection is being followed in 18 states namely Andhra Pradesh, Assam (excluding hill districts). Bihar, Chattisgarh, Goa, Gujarat, Haryana, Himachal Pradesh, Jammu and Kashmir, Jharkhand, Karnataka, Madhya Pradesh, Maharashtra, Punjab, Rajasthan, Tamil Nadu Uttaranchal and Uttar Pradesh and 5 union territories Chandigarh, Dadra and Nagar Haveli, Daman and Due, Delhi and Pondichery.

Permanently Settled States

There are three states namely Kerala, Orissa and West Bengal which come under the category of permanently settled states. In case of these states land revenue was permanently fixed and question of revision ordinarily did not arise. In these states there is no system of recording details of area statistics as there is no permanent revenue staff for a village like village accountant as in the case of temporarily settled area. Initially there was no uniform system of collecting area statistics in these regions.

The police Chaukidar or village headman was usually providing the statistics on the basis of guess work which were however, not very reliable. In order to improve the quality of these statistics in these permanently settled states. A scheme known as "Establishment of Agency for Reporting Agricultural Statistics (EARAS)" was initiated in 1968-69. (Presently, the area statistics in these states are collected by specially appointed field staff under the scheme).

In these States covered by EARAS, the complete enumeration of all fields (survey numbers) *i.e.,* girdawari is conducted every year in a random sample of 20 per cent villages of the States, which are selected in such a way that during a period of 5 years, the entire state is covered. This category covers around 9 per cent of total reported area of the country.

Other Regions

The remaining eight states in North Eastern regions namely Arunachal Pradesh, Manipur, Meghalaya, Mizoram, Nagaland, Sikkim and Tripura and two other union territories namely Andaman and Nicobar Islands and Lakshdweep yet do not have a proper reporting system, though states of Tripura and Skkim (except some minor pockets) are cadastrally surveyed. In these regions compilation of area statistics are based on conventional methods in which estimates are reported by village choukidars on the basis of personal assessment. This category covers around 5 per cent of total reported area of the country.

Non-reporting Areas

Out of the total geographical area of 328 million hectares, land use statistics are available for roughly 306 million hectares. Thus, for about 7 per cent of the geographical area of the country these data are not available. Of the 22 million hectares for which land use data are not available, 17.7 million hectares are located in Jammu and Kashmir and broadly cover the area under illegal occupation of Pakistan and China. The non-reporting area in other States largely consists of hill tracts in Arunachal Pradesh, Nagaland, Manipur and Tripura.

Besides, there are small tracts in some States where due to the absence of cadastral survey and/or the village revenue agency, no regular statistics are collected. Some of these areas are either not accessible or being covered either by forests or by barren mountains.

Regular cadastral survey of these areas is bound to take time. In some areas which are covered by barren hills or which are under snow all round the year, cadastral survey may not also be necessary. For completing the coverage of the land utilisation statistics, it should, however, be possible to estimate the geographical area of these non- reporting areas and their broad land use classification on the basis of aerial photographs coupled with broad topographical survey on the ground. Steps have been taken up with the Governments of Assam, Nagaland, Manipur.

Tripura and Mizoram to prepare ad-hoc estimates of land utilisation for the non-reporting areas falling within their respective territories. In the states where land record are maintained of (temporary settled) the village accountant is in-charge of a village or a group of villages for carrying out field to field crop inspection in each crop season for an agricultural year to record the crop area and land utilisation statistics. He is supposed to record the crop details related to area and land utilisation in Khasra register. After the completion of entries for each survey number of the village a abstract of area sown under different crops "Jinswar statement" is prepared and sent to next higher official in the revenue hierarchy.

At the end of each agricultural year a land utilisation area statistics are compiled and abstract is sent to related higher official. The crop wise and land utilisation wise area statistics obtained from different villages are aggregated at the revenue circle, tehsil and district levels. The district wise area statistics are sent to State Agricultural Statistics

Authority (SASA), which is generally Director of Statistical Bureau or the Director of Agriculture or the Director of Land Records. The state level aggregation is done by SASA and forwarded to Directorate of Economics and Statistics (DES), Ministry of Agriculture and Cooperation, Govt. of India, which is the nodal agency for releasing the state level and the all India level estimates.

Bibliography

A. Rami Horowitz and Isaac Ishaaya.: *Insect Pest Management : Field and Protected Crops*, Springer, London, 2004.

A.K. Dhawan, Balwinder Singh, Manmeet Brar Bhullar and Ramesh Arora.: *Integrated Pest Management*, Scientific Publication, Delhi, 2013.

D.V. Bhagat.: *Encyclopaedia of Insect Pest Management*, Anmol Publication, Delhi, 2010.

Fanindra Prasad Neupane.: *Integrated Pest Management in Nepal : Proceedings of a National Seminar Kathmandu, Nepal 25-26 September 2002*, Himalayan Resources Institute, 2003.

G K Ghosh.: *Biopesticide and Integrated Pest Management*, APH Publication, Delhi, 2009.

H.C.L. Gupta, A.U. Siddiqui and Aruna Parihar.: *Bio Pest Management: Entomopathogenic Nematodes, Microbes and Bioagents*, Agrotech Publication, Delhi, 2010.

Horowitz.: *Insect Pest Management: Field and Protected Crops*, Springer, London, 2003.

K R Nair.: *Integrated Production and Pest Management*, Gene Tech Books, Delhi, 2007.

M L Agarwal.: *Pest Management : A Glossary*, Agrotech Publishing Academy, Delhi, 2007.

N. Venugopal Rao, T. Uma Maheswari, P. Rajendra Prasad, V. Govardhan Naidu and P. Savithri.: *Integrated Insect Pest Management*, Agrobios Publication, Delhi, 2003.

Opender Koul, G.S. Dhaliwal, S.S. Marwaha and Jatinder K. Arora.: *Biopesticides and Pest Management (2 Vols-Set)*, Campus Books International, Delhi, 2003.

P. Narayanasamy, S. Mohan and J.S. Awaknavar.: *Pest Management in Store Grains*, Satish Serial Publication, Delhi, 2009.

Parmeshwar Singh.: *Encyclopaedia of Agricultural Pests and Their Management*, Anmol Publication, Delhi, 2010.

Parvatha Reddy & Abraham Verghese.: *Integrated Pest Management in Horticultural Ecosystems*, Capital Publication, Delhi, 2001.

Pradip V. Jabde.: *Text Book of Applied Zoology : Vermiculture, Apiculture, Sericulture, Lac-Culture, Agricultural Pests and their Control*, Discovery Publishing House, Delhi, 2005.

R. Arora, B. Singh and A.K. Dhawan.: *Theory and Practice of Integrated Pest Management*, Scientific Publication, Delhi, 2012.

R.P. Srivastava.: *Mango Insect Pest Management*, International Publication, Delhi, 1997.

Rajesh Ravi.: *Pest Management : Principles and Practices*, Anmol Publication, Delhi, 2007.

Ram Prakash Srivastava.: *Neem and Pest Management*, IBDCO Publication, Delhi, 2001.

Robert L. Metcalf and William H. Luckmann.: *Introduction to Insect Pest Management*, Wiley , London, 2011.

S Ignacimuthu & Alok Sen.: *Strategies in Integrated Pest Management : Current Trends and Future Prospects*, Phoenix Publication, Nairboi, 2002.

S. Ignacimuthu, s.j. and S. Jayaraj.: *Recent Trends in Insect Pest Management*, Elite Publication, Delhi, 2008.

S.C. Dwivedi and Nalini Dwivedi.: *Integrated Pest Management and Biocontrol*, Pointer Publication, Delhi, 2006.

S.N. Mundra and A.H. Shah.: *Integrated Pest Management in Irrigated Agriculture*, Himanshu Publication, Delhi, 1998.

Shagufta.: *Integrated Pest Management*, APH Publication, Delhi, 2012.

T V Sathe and K P Shinde.: *Dragonflies and Pest Management*, Daya Publication, Delhi, 2008.

T V Sathe, Jyoti M. Oulkar.: *Insect Pest Management: Ecological Concepts*, Daya Publishing House, Delhi, 2010.

T.V. Sathe and A.D. Jadhav.: *Sericulture and Pest Management*, Daya Publication, Delhi, 2001.

Uma Shankar; Satya Priya and Deepak Kumar.: *Vegetable Pest Management Guide for Farmers*, International Book, Delhi, 2008.

Vaishali J. Patil and T.V. Sathe.: *Insect Predators and Pest Management*, Daya Publishing House, Delhi, 2003.

Vikas Chaudhary.: *Entomology and Pest Management* , Navyug Publication, Delhi, 2008.

Index